Analysis of Enzyme Kinetic Data

Analysis of Enzyme Kinetic Data

ATHEL CORNISH-BOWDEN

*Directeur de Recherche, Centre National de la
Recherche Scientifique, Marseilles*

Oxford   New York   Tokyo

OXFORD UNIVERSITY PRESS

1995

Oxford University Press, Walton Street, Oxford OX2 6DP
Oxford   New York
Athens   Auckland   Bangkok   Bombay
Calcutta   Cape Town   Dar es Salaam   Delhi
Florence   Hong Kong   Istanbul   Karachi
Kuala Lumpur   Madras   Madrid   Melbourne
Mexico City   Nairobi   Paris   Singapore
Taipai   Tokyo   Toronto
and associated companies in
Berlin   Ibadan

Oxford is a trade mark of Oxford University Press

Published in the United States
by Oxford University Press Inc., New York

© Oxford University Press, 1995

A catalogue record for this book is available from the British Library

Library of Congress Cataloging in Publication Data
Cornish-Bowden, Athel.
Analysis of enzyme kinetic data / Athel Cornish-Bowden.
Includes bibliographical references and index.
1. Enzyme kinetics—Statistical methods. 2. Enzyme kinetics—Data
processing. I. Title.
[DNLM: 1. Enzymes—pharmacokinetics. 2. Regression Analysis.
3. Software. -qU 135 C818a 1995]
QP601.3.C67   1994   574.19'25—dc20   94-32907

ISBN 0 19 854878 8 (Hbk)
ISBN 0 19 854877 X (Pbk)

Typeset by Technical Typesetting Ireland
Printed in Great Britain by Biddles Ltd., Guildford & King's Lynn

To Marilú

Preface

This is a book that I have wanted to write (the first six chapters, at least) for many years, and, indeed, I made a start on an ancestral version during a sabbatical in 1977. However, it soon became clear that a short book on the theory of data analysis in enzymology would have very limited appeal, and for this and various other reasons the original project did not advance very far. The arrival of the personal computer has completely transformed the world of scientific computing, however, to the point where virtually every working scientist is now also a computer user. As a result, it has become quite feasible to incorporate all of the methods of analysis developed in the 1960s and 1970s into a single program and to present both the methods and the program in a single book.

The two principal parts of the book are largely independent of one another, with only a short link section (Chapter 7) between them: the first six chapters provide a theoretical account of statistical analysis of kinetic data for enzyme-catalysed reactions in the steady state; the last four describe Leonora, a program for analysing enzyme kinetic data on the IBM PC and compatible computers. Each of these parts can be read almost independently of the other, and each makes very little reference to the other. One may reasonably ask, therefore, why they have been bound between the same pair of covers and offered as a single book. The answer is that although they can be read in isolation from one another, that is far from being the best way to proceed.

Something that will strike anyone who pays more than passing attention to statistics journals is that the number of statistical methods that have been proposed for scientists and engineers to use is much larger than the number of such methods that are actually in use by scientists or engineers. This is less true of methods proposed in journals that are normally read by their potential users, but it is still true to some degree of methods of kinetic data analysis that one can find in the biochemical literature. It is one thing to be reasonably convinced by a research article that a new method is better than existing ones, but it is quite another to go out and use it in the laboratory if one has to develop it from nothing. The reason for adding the practical part of this book (and the accompanying software) to the theoretical part, therefore, is to provide the tools necessary for the reader to test and apply all of the theory.

This leaves unanswered, however, the complementary question of why the user of the software would want to be bothered with the theory. The reason is in the sort of program that Leonora is. It does not follow the philosophy of assuming that there is One True Way of analysing data that must be applied in all circumstances. On the contrary, it offers a great deal of choice to its users, though to avoid making use too difficult it makes its own (i.e. my) choices when others are not made. To make appropriate choices the user needs a theoretical point of reference. Moreover, when I started writing Leonora I intended it to permit use of virtually any method the user might wish to try, but in practice the number of possibilities is almost infinite, because the more choices allowed, the more sub-choices these imply, etc. Consequently Leonora does make some restrictions, but to know why these restrictions apply rather than others one again needs a point of reference.

Not all of the methods Leonora offers are in my opinion good methods, and even if they were it would be reasonable to ask what is the point of offering so many. This comes back to the question of user choice: far too many programs of all kinds are written in the spirit of the One True Way, and when they tolerate different preferences from those of the programmer they may force them to be specified every time the program is used. In the case of enzyme kinetics some of the most widely used methods come into the category of bad methods, but the solution is not to forbid their use—if potential users do not find the methods they regard as most natural they will not continue to be users—but to try to persuade users that better methods exist that are just as convenient. The ideal, in my view, is not only to offer a choice, but also to offer users who do not want to avail themselves of the choice a default method that will work well in most circumstances.

Anyone using Leonora to analyse results of research experiments will, in all likelihood, settle quite soon on one method of analysis and ignore the others. Not everyone is a researcher, however, and for teaching the principles of data analysis there is a more obvious need for a program that will allow the use and comparison of many different methods. Leonora is intended to address this need.

I am grateful to the Faculty of Sciences of the University of Chile for appointing me on two occasions to the visiting Chair set up in memory of the late Professor Hermann Niemeyer Fernández, and to the members of the Laboratory of Biochemistry in Santiago for providing me with the opportunity to do much of the work on this book there. Work on some of the methods described in the book benefitted greatly from collaborations with Robert Eisenthal and Laszlo Endrenyi, and I thank both of them for this. I thank Véronique Raphel for allowing me to use data from her doctoral thesis as the practical example around which Chapter 7 is written. Finally, I thank Marilú Cárdenas for a great deal of help in checking the proofs, without which many errors would have escaped detection.

Marseilles   September 1994   A. C-B.

Contents

I Theory

II   Interlude

III   Practice

I Theory

1
Least-squares analysis: basic principles

1.1 The statistical approach to data analysis

'Go on, Mr. Pratt,' says Mrs. Sampson, 'Them ideas is so original and soothing. I think statistics are just as lovely as they can be.'

O. Henry, *The handbook of Hymen*

It is probably still true that most experiments in biochemical kinetics are analysed by graphical methods, though the appearance in the 1970s of cheap electronic calculators with built-in functions for linear regression increased the frequency with which some sort of a pretence to a statistical analysis is used, a tendency that has continued now that computers are to be found in every laboratory. Nonetheless, since the landmark papers of Wilkinson (1961) and Johansen and Lumry (1961), which were brought to a wider audience by a review of statistical methods by Cleland (1967), it has been increasingly recognized that satisfactory analysis of kinetic data requires even the best graphical methods to be supplemented by more detailed statistical calculations. In any case, to appreciate why different ways of plotting the same data do not necessarily lead to the same conclusions, and why some plots are better than others, one needs to have some understanding of the statistical implications of using particular plots.

If all experiments were carried out with perfect accuracy on perfectly uniform and consistent materials there would be no need for statistical analysis and all algebraically valid graphs and calculations would give the same result. In the real world, however, one must deal with biological samples with variable properties and make measurements on them that are subject to experimental error. In enzyme kinetics, variations between different preparations of enzyme normally make it impractical to carry out detailed analysis with results that have been combined from experiments made over a long period with different preparations. Indeed, most experimenters prefer not to combine results from different days, except qualitatively, unless there is good evidence that the preparation shows no changes in properties over an extended period, or only changes that can be fully corrected for. Experimental errors in the measurements remain, however, and can never be wholly eliminated. The object of statistical analysis, therefore, is both to minimize the effect of experimental error on the conclusions, and also to assess how large the residual effect is, i.e. to estimate the precision of the results.

The classical approach to statistical analysis, the only approach that is considered in most elementary treatments, is based either explicitly or implicitly on an assumption that the laws governing the kinds of errors that arise in experiments are either known or can easily be determined. If this is true, one can use the known distribution of errors to determine the type of analysis that will lead to the highest possible accuracy in the conclusions. This approach, which is known as *parametric statistics* because the analysis is based on the parameters of the distribution curve, will be discussed in the first chapters of this book. In Chapter 4 I shall discuss whether the assumptions underlying classical methods can really be trusted, and in Chapters 5 and 6 I shall describe alternative methods that are less dependent on assumptions. However, because the classical approach is adequate for many purposes and anyway needs to be understood as a starting point for considering alternatives, I shall reserve any scepticism about it for the moment.

Some readers may question whether it is necessary to consider statistical methods in as much detail as will be found in this book, feeling that everything that an experimenter needs to know can be found in textbooks of statistics or in the documentation accompanying computer programs. However, the more elementary statistics textbooks tend to be largely devoted to experiments rather different from those carried out in kinetics. In particular, virtually all of the models needed for kinetic studies are non-linear, whereas if elementary textbooks discuss model fitting at all they normally confine discussion to linear models. Even books devoted to regression techniques commonly treat non-linear problems as an afterthought. The more advanced books are, moreover, rendered difficult for the non-specialist to use by their enthusiastic use of matrix algebra: although this provides an elegant and highly concise way of expressing rather complicated relationships, it is at the expense, for people not well practised in matrix calculations, of a loss in comprehensibility that is almost total. In this book I shall keep matrix algebra to a minimum.

1.2   The continuing importance of graphs

There is no single statistical tool that is as powerful as a well-chosen graph. Our eye–brain system is the most sophisticated information processor ever developed, and through graphical displays we can put this system to good use to obtain deep insight into the structure of data.

Chambers et al. 1983

Far better an approximate answer to the *right* question, which is often vague, than an *exact* answer to the wrong question, which can always be made precise.

Tukey 1962

It has sometimes been suggested (e.g. Cornish-Bowden 1976) that statistical calculations make graphical analysis unnecessary, and the increasing availability of computer packages for fitting non-linear models to experimental data has certainly reinforced this idea. It is, however, mistaken, because graphs and statistical calculations have their own strengths and weaknesses, and should be regarded as complementing one another rather than as alternatives. Moreover, different types of graph also complement one another, and no single type is optimal for all purposes. To take a familiar example, the direct linear plot (Eisenthal and Cornish-Bowden 1974) provides a satisfactory method of estimating the parameters of the Michaelis–Menten equation when this equation is known to be obeyed, but it is less effective as an aid to recognizing when it is not obeyed; conversely, the direct plot of rate against substrate concentration is almost useless as a means of estimating the parameters, but is among the better plots for illustrating deviations from the expected behaviour.

The continuing need for good graphical methods, as well as quick and easy (but not necessarily precise) methods of calculation, has been stressed in particular by Tukey, who has also been responsible for many of the most imaginative and important advances in statistical analysis. In his book *Exploratory data analysis* (Tukey 1977) the emphasis throughout is on the closest possible contact between the analyst and the data, often plotting or displaying the same data in several different ways to look for particular kinds of information in it. Nonetheless, one should avoid the opposite error of thinking that detailed analysis will extract meaning from meaningless experiments, and Tukey is also, I believe, responsible for the aphorism that

If a thing is not worth doing at all, it is not worth doing *well!*

This should underlie the thoughtful experimenter's attitude to statistical calculations. It is pointless to devote time and effort to very detailed and rigorous analysis of data that are simply not of a high enough standard to be worth more than a cursory glance. The effort would be better directed towards careful experimental design, including attention to better experimental techniques, to obtain data capable of yielding the sort of conclusions wanted. Even the best statistical calculations cannot extract information from experiments if it is not there to be extracted.

Why are graphs important to this discussion? What types of information are more easily obtained from graphs than from calculations? In general, the human eye is very good at recognizing unexpected behaviour in a graph, but much less good at recognizing it in columns of numbers in a table. For example, if the results of fitting a model to a set of kinetic measurements yield results in which all of the calculated rates below a certain substrate concentration are less than the observed rates, whereas all of the others are greater, this ought to be obvious from a table if the calculated differences are printed (because of the systematic arrangement of signs). It may be less obvious if they are not printed, or if the observations are not arranged in increasing or decreasing order of concentration; and more subtle deviations from the model may be much harder to see, especially if they are accompanied by a substantial amount of random error. However, the same deviations will often be very easy to see if the results are plotted, especially if a residual plot (Sections 3.4–3.5) is used. Even if the plot does not make it clear whether the systematic effects are big enough to be accepted definitely as genuine, it will normally make it clear that more detailed examination is needed.

The results of fitting a model to data should always be plotted, preferably in more than one way, before the model is accepted as satisfactory, even if no graphs are needed for publication, etc. The fact that a good computer program may have included some tests of whether the fit is satisfactory does not remove the need for plotting. Computer programs are more easily deceived than the eye, and in any case they can only test for the sorts of deviation that the programmer has considered as possibilities. Thus a test for linear deviations may fail to detect quite obvious quadratic deviations, and a test for quadratic deviations may fail to detect quite obvious higher-order deviations, etc. The eye, by contrast, can easily notice behaviour that has not been reported or observed before, such as the effect of rounding error on a residual plot (Section 3.5).

Table 1.1   Michaelis–Menten data

The Table shows five sets of substrate concentrations a and rates v, together with calculated rates $\hat{v}$. The values of a and $\hat{v}$ are identical for data sets (a) to (d), but are different for set (e). All five sets of data generate the same best-fit Michaelis–Menten parameters when analysed according to Section 2.3 (eqns 2.12 and 2.13) and Section 2.4, i.e. $\hat{K}_m = 2.7 \pm 0.21$, $\hat{V} = 18.2 \pm 0.67$, which were used for calculating the $\hat{v}$ values shown.

Sets (a) to (d)		Set (a)	Set (b)	Set (c)	Set (d)	Set (e)		
a	$\hat{v}$	v	v	v	v	a	$\hat{v}$	v
12.0	14.9	13.3	17.5	15.4	15.0	2.85	9.35	9.32
1.5	6.50	6.4	6.45	6.23	6.0	2.85	9.35	9.01
6.0	12.6	13.3	12.2	13.1	12.0	2.85	9.35	9.12
15.0	15.4	16.2	14.9	15.2	15.0	2.85	9.35	9.30
0.5	2.84	2.95	2.92	3.01	3.0	2.85	9.35	9.66
8.0	13.6	14.4	13.2	14.4	15.0	2.85	9.35	9.49
2.0	7.74	7.91	7.67	7.52	7.0	2.85	9.35	9.58
20.0	16.0	15.9	15.4	13.7	15.0	12.0	14.9	14.9
10.0	14.3	15.0	13.8	15.1	15.0	2.85	9.35	9.34
5.0	11.8	11.2	11.5	12.2	12.0	2.85	9.35	9.56
1.0	4.92	4.94	4.97	4.67	5.0	2.85	9.35	9.37
3.0	9.58	8.45	9.39	9.54	10.0	2.85	9.35	9.11

Table 1.1 illustrates these points. All of the five data sets shown contain the same number of observations and lead to the same estimates of the Michaelis–Menten parameters, with the same precision estimates, and one might well suppose that they represented equally good experiments and that there was no reason to prefer any to any other. Just studying the numbers in the table makes it quite difficult to see any important differences between them. However, as soon as one plots the data in the form of residual plots, the differences leap out at the eye (Fig. 1.1)

Only data set (a) can now be called satisfactory, with the points in the residual plot arranged as one would expect for a random scatter. Each of the others illustrates some anomaly in the data, shown here in an extreme form; milder forms of the same anomalies occur frequently in real examples of data analysis. Data set (b) illustrates the effect of one very bad observation: if this is the consequence of a typing error when entering the data it can be corrected, or if it can be shown to result from a mistake in the execution of the experiment it can be removed from the analysis, and in either case much more precise parameter values can then be obtained. If neither of these is the cause there is clearly a peculiarity that needs further study. Data set (c) shows obvious evidence of systematic error: the points fall on a well-defined line, but the Michaelis–Menten equation is not the right model. Data set (d) shows the effect of over-aggressive rounding, to the extent that the residuals are principally determined by rounding error: the remedy is obvious, but it can only be applied if the problem is noticed. Data set (e) is the only one for which the anomaly is clearly visible in Table 1.1, but it is even more obvious in the residual plot: it is an example of very poor experimental design, in which only two different substrate concentrations have been used, with all but one of the observations at the same concentration. With such a design the parameter estimates are heavily dependent on the correctness of the isolated observation.

This is, of course, an artificially constructed example, adapted from one discussed by Anscombe (1973) in the context of straight-line regression. More realistic, less extreme, examples make the case for graphical analysis stronger, not weaker, because in such examples it becomes even more difficult to perceive any anomalies in the numerical information although it remains quite easy to recognize them in a graph. One might hope that a well-designed computer program would incorporate tests to recognize anomalies in the data, but this hope is far from being a reality at present. I know of no existing computer program capable of recognizing any anomaly in data set (d), and there are many that would notice nothing odd about data sets (b), (c), or (e).

Although these sorts of problem are most evident in a residual plot, even an ordinary plot of dependent variable against independent variable can often reveal information that is not obvious in a table. With only a table of numbers one might well accept that the model represented by the line in

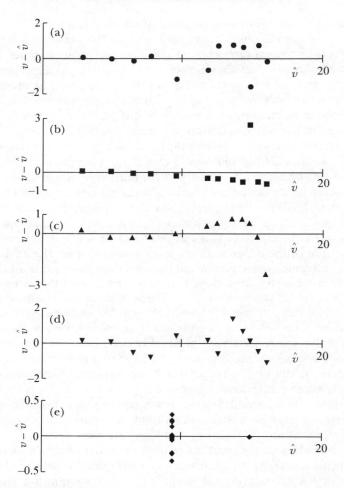

Fig. 1.1. *Residual plots from 'equivalent' sets of data.* Values of $v - \hat{v}$ are plotted against $\hat{v}$ for the five sets of data listed in Table 1.1 (a) A random scatter of points illustrates the ideal case, with no abnormality; (b) one point far off the line for the others suggests a typing error or a gross experimental error, but a less extreme example could result from an unusual property that merits further investigation; (c) a systematic arrangement of points suggests that the wrong model has been fitted; (d) several disconnected systematic regions suggest that the data were excessively rounded before analysis; (e) a very poor experimental design with almost all the information obtained from a single substrate concentration produces results that are heavily dependent on the accuracy of the isolated point, and with only two concentrations represented it is impossible to know whether any systematic error is present.

Fig. 1.2 gave an adequate fit to the data, but the graph shows that significant curvature in the data is not taken account of by the line; a

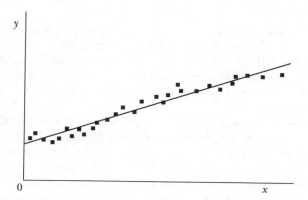

Fig. 1.2. *Visual impact of a graph.* In this graph it is immediately evident that the points do not lie well on the straight line, and that a line with a slight downward curvature would fit them much better. Near the y axis there is a suggestion of curvature in the opposite direction, but establishing this would need additional observations. None of this would be evident from a table of (x, y) values accompanied by the equation for the straight line, and would be difficult to recognize even if the table included calculated y values.

residual plot of the same data would make the curvature too obvious to be ignored.

Good computer programs often include the facility of producing plots of the results automatically, and these may sometimes dispense with the need to make plots manually. This is especially true if the program is able to drive a high-quality plotter, or if the plots are residual plots (for which exact plotting is not needed), but it will be rare that no manual intervention is needed. For example, if the program is given no guidance from the experimenter it will be unable to make the best decisions about which data should be plotted, over which range, and with which models shown together on the same plot.

1.3   True values, population values, observed values, and estimates

When a quantity is measured experimentally, the resulting *observed value* will in general differ, because of experimental error, from the true value, which may itself differ from the ideal value because of sampling variation, and statistical calculations will yield a *best-fit estimate* that will in general differ from all of the other three. We need, therefore, to consider why the differences arise, how their magnitude can be assessed, and how we can symbolize the different kinds of values to make it clear which we are referring to.

For a physical quantity, such as the atomic mass of chlorine, sampling variation is likely to be non-existent or at least trivial in comparison with experimental error: one atom of ^{35}Cl is, we believe, completely identical with another, and although ^{35}Cl occurs in nature mixed with ^{37}Cl the mixture is normally completely uniform and consistent. Thus one sample of chlorine should lead to exactly the same atomic mass as another, and if any sampling variation did arise (for example because of varying amounts of impurities) we would probably regard it as experimental error rather than as sampling variation.

For biological quantities, however, the situation is quite different. If several determinations of the rate of glucose phosphorylation in perfused rat liver yielded different values we would assume that only a part of the difference, probably a small part, was a consequence of experimental error. The main variation would be ascribed to the fact that not every rat liver is identical with every other, and even when measurements are made on samples from one animal or one organ, or one sample at different times, we would assume that there are some real variations in properties. Accordingly, therefore, we regard the properties of the sample as approximations to the properties that are representative of the entire population of rats. We suppose that if identical measurements were made on identically prepared samples from all of the rats that have ever existed then the values would be distributed about a mean value. This rather idealized value is known as the *population mean* or the *population value* of the quantity of interest.

Even if we are working with a homogeneous preparation of a biological material, such as a purified enzyme, and even if we believe that there is no sampling variation in sub-samples taken from this preparation, the idea of a population is still relevant. Each measurement that we make is subject to experimental error, and can be regarded as a sample from the infinite population of measurements that we might have made. When experimental error is the primary source of variation, however, we often prefer to regard the population value of the parameter as the *true value*, though the degree of 'truth' may vary with the parameter in question. Thus of the parameters of the Michaelis–Menten equation, the limiting rate V will certainly vary from one preparation to another, but the Michaelis constant K_m may vary much less. This contrasts, incidentally, with our capacity to estimate these two parameters in any given experiment, where K_m is always subject to proportionately greater errors than V.

The most important point to note about the true or population value of a parameter is that it is always unknown, and remains unknown regardless of how many measurements are made. Consequently it cannot be used in calculations, and should be symbolized in a way that makes it clearly distinct from quantities with known values. In theoretical statistics population values are often symbolized by greek letters, such as β for the true value of a quantity whose observed value is b. This convention will be used as far as possible in this book, though the lack of greek equivalents to

v and V, and of distinguishable greek equivalents to k and K makes it a less than ideal convention for use in kinetics.

The observed values of physical quantities can be regarded as estimates (not necessarily best estimates) of the corresponding true or population values, and are conventionally represented by roman letters. Estimates of parameters that are not directly observed, such as the slope of a straight line, are represented in the same way. *Best-fit estimates*, such as those that satisfy the *least-squares* criterion, are conventionally distinguished from observed values and other estimates by a circumflex accent. Thus $\hat{b}$ (spoken aloud as 'b-hat') is the best-fit estimate corresponding to β. Estimates that are best by some other criterion (such as the *robust* estimates to be discussed in Chapter 6) are also sometimes indicated by a circumflex accent, but it is advantageous to use a distinctive symbol to avoid confusion, and in this book such estimates will be written with asterisks, e.g. b^* ('b-star').

1.4 Variance

As we have considered, the population value of a measured quantity b is the value approached by the sample mean as the sample size approaches infinity, i.e.

$$\mu = \lim_{n \to \infty} \frac{1}{n} \sum_{i=1}^{n} b_i. \tag{1.1}$$

(Although the summation limits are shown explicitly in this equation, from 1 to n, these are usually obvious in statistical calculations, i.e. summations are carried out over the whole sample or the whole data set. Accordingly, limits will not be explicitly shown in the remainder of this book.) If b is an estimate of a quantity with true value β, then it is, by definition, an *unbiassed* estimate if the population mean μ defined by eqn. 1.1 is identical with β.

It follows from this definition that the mean of $b_i - \beta$ approaches zero as n approaches infinity if b_i is unbiassed, and thus does not provide a useful measure of the variability of b_i. Such a measure is, however, provided by the mean *squared* difference between the estimates and the true value, which does not approach zero because every value is positive. This quantity is known as the *variance* of b_i, $\sigma^2(b_i)$, and is defined by the following equation:

$$\sigma^2(b_i) = \lim_{n \to \infty} \frac{1}{n} \sum (b_i - \beta)^2. \tag{1.2}$$

As β is unknown and real sample sizes are always finite, the variance as defined by eqn 1.2 can never be measured, but it can be estimated from the *sample variance* $s^2(b_i)$ defined as follows:

$$s^2(b_i) = \frac{1}{n} \sum (b_i - \hat{b})^2 \qquad (1.3)$$

where n is now finite and $\hat{b}$ is a best-fit estimate of β. This immediately suggests a way in which we may define 'best-fit': we can take the best-fit estimate $\hat{b}$ of β as the value that makes the expression for the sample variance a minimum; then a simple calculation shows that $\hat{b}$ is identical with the sample mean $\overline{b}$, i.e.

$$\hat{b} = \overline{b} = \frac{1}{n} \sum b_i. \qquad (1.4)$$

The sample variance is actually a *biassed* estimate of the population variance defined by eqn 1.2. This is because $\hat{b}$ is *not* the true value β, and of all the ways we might have defined it we have chosen the one that makes the sample variance a minimum, so we should expect that on average the sample variance will be smaller than the population variance. After we have obtained an expression for the variance of $\hat{b}$, eqn 1.11 below, we can estimate the magnitude of the bias.

First we shall need an expression for the variance of the sum of two numbers $(x + y)$, where x and y have population means μ_x and μ_y respectively, and variances $\sigma^2(x)$ and $\sigma^2(y)$ respectively. The mean of the sum is clearly $\mu_x + \mu_y$, and according to eqn 1.2 the variance must be the population mean of all terms of the type

$$(x_i + y_i - \mu_x - \mu_y)^2 = (x_i - \mu_x)^2 + 2(x_i - \mu_x)(y_i - \mu_y) + (y_i - \mu_y)^2. \qquad (1.5)$$

It must therefore be given by

$$\sigma^2(x + y) = \sigma^2(x) + 2\,\mathrm{cov}(x, y) + \sigma^2(y), \qquad (1.6)$$

where $\mathrm{cov}(x, y)$ is a new quantity called the *covariance* of x and y and is defined as follows:

$$\mathrm{cov}(x, y) = \lim_{n \to \infty} \frac{1}{n} \sum (x_i - \mu_x)(y_i - \mu_y). \qquad (1.7)$$

It is evident from this definition that the covariance is a measure of the tendency of x and y to vary in the same way. If x and y are *independent*, it is zero; indeed, this is the definition of independence in a statistical context. However, if there is some source of error that tends to affect x and y in the same direction the covariance will be positive, whereas if it affects them in opposite directions it is negative. If x and y are independent, however, eqn 1.6 simplifies to the following:

$$\sigma^2(x + y) = \sigma^2(x) + \sigma^2(y). \tag{1.8}$$

This result then defines the important additive property of the variance for sums of independent variables. The same expression defines the variance of the difference $(x - y)$ between independent variables, because the variance of $-y$ is the same as that of y:

$$\sigma^2(x - y) = \sigma^2(x) + \sigma^2(y). \tag{1.9}$$

Equations 1.8 and 1.9 are useful, but one should avoid becoming so familiar with them that one forgets that they are *not* general expressions for the variance of a sum or a difference, but are only special cases of the proper relationship given by eqn 1.6.

The variance of the product nb of an exact constant n and an estimate b follows in an obvious way from eqn 1.2:

$$\sigma^2(nb) = n^2\sigma^2(b). \tag{1.10}$$

Taken with eqn 1.8 this leads straightforwardly to an expression for the best-fit estimate $\hat{b}$ defined by eqn 1.4, provided that the individual b_i values are independent:

$$\sigma^2(\hat{b}) = \frac{1}{n^2}\left[\sigma^2(b_1) + \sigma^2(b_2) + \cdots + \sigma^2(b_n)\right], \tag{1.11}$$

which simplifies to

$$\sigma^2(\hat{b}) = \frac{1}{n}\sigma^2(b) \tag{1.12}$$

if every b_i has the same variance. (The case where every b_i does not have the same variance is discussed in Section 1.5.)

This result allows us to assess the extent of the bias in the sample variance $s^2(b)$ as an estimate of the population variance $\sigma^2(b)$. As mentioned above, this bias arises from the fact that $s^2(b)$ is calculated from the deviations of the b_i from $\hat{b}$, whereas an unbiassed estimate of $\sigma^2(b)$ should be based on deviations from β. However,

$$(b - \beta) = (b - \hat{b}) + (\hat{b} - \beta), \qquad (1.13)$$

and so

$$(b - \beta)^2 = (b - \hat{b})^2 + (\hat{b} - \beta)^2 + 2(b - \hat{b})(\hat{b} - \beta). \qquad (1.14)$$

Examining the third term on the right-hand side, we note that $(\hat{b} - \beta)$ is a constant and that $(b - \hat{b})$ has a mean of zero, and so we can write

$$E\left[(b - \beta)^2\right] = E\left[(b - \hat{b})^2\right] + E\left[(\hat{b} - \beta)^2\right], \qquad (1.15)$$

where the notation E( ) represents the *expected value* of an estimate, defined as the mean of its distribution. As the expected value of a squared deviation from the mean is, by definition, the variance, this is equivalent to

$$\sigma^2(b) = E[\,s^2(b)\,] + \sigma^2(\hat{b}) = E[\,s^2(b)\,] + \frac{1}{n}\,\sigma^2(b). \qquad (1.16)$$

Hence,

$$E[\,s^2(b)\,] = \left(\frac{n-1}{n}\right)\sigma^2(b) \qquad (1.17)$$

and so $s^2(b)$ can be converted into an unbiassed estimate of $\sigma^2(b)$ by multiplying it by $n/(n-1)$.

It often happens in fitting a model to experimental data that a parameter of interest is the product or ratio of two parameters whose variances can be estimated directly, so it is important to have an expression for the variance of a product xy of two estimates x and y. This may be found by writing x as $\mu_x + \varepsilon_x$, i.e. the sum of its population mean and an error, and y similarly as $\mu_y + \varepsilon_y$, so that

$$xy = (\mu_x + \varepsilon_x)(\mu_y + \varepsilon_y) = \mu_x\mu_y + \mu_x\varepsilon_y + \mu_y\varepsilon_x + \varepsilon_x\varepsilon_y. \qquad (1.18)$$

As the first term is exact its variance is zero, and so provided that the variances are small enough for the last term to be negligible, we need consider only the middle two terms in relation to eqn 1.6, the expression for the variance of a sum, to write down an expression for the variance of xy:

$$\sigma^2(xy) \approx \hat{y}^2\sigma^2(x) + 2\,\hat{x}\hat{y}\,\text{cov}(x, y) + \hat{x}^2\sigma^2(y), \qquad (1.19)$$

where the unknown population means have been replaced by best-fit estimates $\hat{x}$ and $\hat{y}$. Note that for three separate reasons this result is not exact: first, it is valid to ignore the last term in eqn 1.18 only if the variances are small; second, the true variance depends on the population means, not on estimates of them; third, even if $\hat{x}$ and $\hat{y}$ are unbiassed estimates of these means, $\hat{x}^2$ and $\hat{y}^2$ are *not* unbiassed estimates of μ_x^2 and μ_y^2. This last point is perhaps surprising, but it illustrates an important general principle that must be understood if serious mistakes are to be avoided when the variances are large: if b is an unbiassed estimate of β, $f(b)$ is not an unbiassed estimate of $f(\beta)$, where $f(\ )$ is any non-linear function. It is also important to note that although eqn 1.9 may sometimes be an acceptable approximation to eqn 1.6, the parameters that are multiplied together to produce other parameters are virtually never independent, and consequently the middle term must not be omitted from eqn 1.19.

The expression for the variance of a ratio x/y may be derived by similar arguments, and is as follows:

$$\sigma^2\left(\frac{x}{y}\right) \approx \frac{\sigma^2(x)}{\hat{y}^2} - \frac{2\,\hat{x}\,\mathrm{cov}(x, y)}{\hat{y}^3} + \frac{\hat{x}^2\sigma^2(y)}{\hat{y}^4}. \tag{1.20}$$

For the same reasons as above, this expression is not exact, and becomes seriously incorrect as the errors in y exceed about 10 per cent.

Equations 1.19 and 1.20 may be written in a slightly different way that makes their relationship to one another more obvious, and hence allows them to be remembered more easily:

$$\sigma^2(xy) \approx \hat{x}^2\hat{y}^2\left[\frac{\sigma^2(x)}{\hat{x}^2} + \frac{2\,\mathrm{cov}(x, y)}{\hat{x}\hat{y}} + \frac{\sigma^2(y)}{\hat{y}^2}\right], \tag{1.21}$$

$$\sigma^2\left(\frac{x}{y}\right) \approx \frac{\hat{x}^2}{\hat{y}^2}\left[\frac{\sigma^2(x)}{\hat{x}^2} - \frac{2\,\mathrm{cov}(x, y)}{\hat{x}\hat{y}} + \frac{\sigma^2(y)}{\hat{y}^2}\right]. \tag{1.22}$$

If x is exactly equal to 1 with zero variance, eqn 1.20 leads directly to an expression for the variance of a reciprocal, which is useful for examining the properties of the linear plots often used in enzyme kinetics:

$$\sigma^2(1/y) \approx \frac{\sigma^2(y)}{\hat{y}^4}. \tag{1.23}$$

Although the variance has algebraic properties that make it particularly valuable for theoretical discussions, it is less convenient for applying to experimental results, because its dimensions and units are different from those of the variable to which it refers. In practice, therefore, the results of statistical calculations are often reported in terms of the square root of the variance, σ or s. When σ or s refers to a set of values that have been observed directly, it is called the *standard deviation*, but when it refers to a derived quantity, such as a mean or a best-fit parameter, it is called the *standard error*. It is also convenient to be able to express precision in dimensionless terms (as a per cent or as a ratio), and the standard deviation (or standard error) divided by the population value or a best-fit estimate of it is called the *coefficient of variation*.

1.5   Weighting

Elementary accounts of statistical methods commonly give little attention to the possibility that the experimental values analysed may not all be equally reliable, but this is a simplification that we cannot afford in enzyme kinetics. Although it may sometimes be true (albeit less often than is commonly supposed) that the primary observations, a set of v values for example, have uniform variance, we may need to transform them before analysis into secondary values, such as $1/v$, that are certainly not of uniform variance. We must therefore consider whether $\hat{b}$ as defined by eqn 1.4 remains a best-fit estimate of β in the event that the individual b_i values have various different variances.

When a set of observations are of uniform variance, they are sometimes said to be *homoscedastic*, and when they are not they are *heteroscedastic*, and the corresponding properties are called *homoscedasticity* and *heteroscedasticity* respectively. We shall not need any of these terms in this chapter (though they could be used in numerous places if one wanted to be obscure), but they are introduced here because they will be useful in Chapter 6 when we come to consider ways of assessing the variations in variance within sets of observations.

Suppose that in a set of n values b_i each has a known variance σ_i^2. Intuitively, we should expect that in estimating β we should allow for this variation of variances by calculating a *weighted* mean,

$$\hat{b} = \sum w_i b_i \Big/ \sum w_i, \qquad (1.24)$$

in which the values of the weights w_i are larger for the more reliable observations. However, intuition provides no guide to how much larger they ought to be, i.e. it does not immediately suggest what function of σ_i^2 ought to be used for calculating w_i. This information comes from the fact that if the individual b_i values are independent, the variance of $\hat{b}$ as defined by eqn 1.24 follows from eqn 1.8:

$$\sigma^2(\hat{b}) = \sum w_i^2 \sigma_i^2 \Big/ \left(\sum w_i \right)^2, \qquad (1.25)$$

and partially differentiating with respect to any individual weight w_j we can readily show that the partial derivative is zero if

$$w_j = \sum w_i^2 \sigma_i^2 \Big/ \sigma_j^2 \left(\sum w_i \right)^2, \qquad (1.26)$$

and it is obvious from inspection that it is only possible for *all* such partial derivatives to be simultaneously zero if every weight is inversely proportional to the corresponding variance, i.e.

$$w_i = 1/\sigma_i^2. \qquad (1.27)$$

The constant of proportionality, shown here as 1, is arbitrary and can have any value (apart from zero) without affecting the value of $\hat{b}$ or its variance. It is often convenient to choose a constant such that the average w_i is about 1.

Weights defined in accordance with eqn 1.27 are appropriate in any least-squares calculation when the variances of the observations are known or when the function that defines how they vary is known. In reality one normally does not know with confidence how the variances of the observations vary, but if a reasonable assumption is possible eqn 1.27 can still be used to calculate suitable weights. In enzyme kinetics one often measures a rate v at substrate concentration a but transforms v into $1/v$ or a/v before plotting, and the question then arises as to how the transformation affects the weighting. From eqns 1.23 and 1.10 it is evident that the variances of $1/v$ and a/v must be as follows:

$$\sigma^2(1/v) \approx v^4 \sigma^2(v), \qquad (1.28)$$

$$\sigma^2(a/v) \approx \left(\frac{v^4}{a^2} \right) \sigma^2(v), \qquad (1.29)$$

where $\sigma^2(v)$ is the variance of v and the concentration a is assumed to be known without error. (If errors in a are significant then a more complex expression similar to eqn 1.20 is needed.) Provided the weights appropriate for the v values are known, therefore, the appropriate weights for $1/v$ and a/v can readily be calculated. In practice two hypotheses for the appropriate weights for v are commonly invoked: (i) that the v values are of *uniform variance*, so that

$$w(1/v) = v^4, \qquad (1.30)$$

$$w(a/v) = v^4/a^2, \qquad (1.31)$$

or (ii) that the v values have *uniform coefficient of variation*, i.e. that their variances are proportional to the squares of the true v values, in which case

$$w(1/v) = v^2, \tag{1.32}$$

$$w(a/v) = v^2/a^2. \tag{1.33}$$

All of the expressions in eqns 1.30–1.33 are approximate, because they are derived from approximate expressions for the variances, but they are written with equality signs because if they are to be used in calculations actual values must be put to the weights even if there is no exact way of calculating them. In any case, the errors that result from these approximations are usually trivial in comparison with the errors that result from using an inappropriate expression, such as putting all weights for $1/v$ to unity rather than using eqn 1.30. Moreover, the weights implied by the two kinds of assumption are usually very different from one another and lead to accordingly different final results in any calculation in which they are used. These points will be illustrated by a numerical example in the next chapter (Table 2.1). It follows then, that it is of some importance to know which (if either) of the simple hypotheses ought to be invoked, and ways of deciding this will be discussed in Section 6.3.

Cleland (1979) regards the choice of weighting assumption as no more than a matter of the range of v values—if they span less than a factor of 5 the variance is constant, but if they span more than a factor of 10 the coefficient of variation is constant (with, presumably, something intermediate if they span a factor between 5 and 10). However, a moment's reflection will show that this cannot be correct: how can the variances of the rates already measured depend on whether you decide to make a last measurement at a very low concentration before going home? Remember that making the wrong choice with rates spanning more than a factor of 10 means setting weights that are wrong by more than a factor of 100, so it can be an expensive mistake to make. Nonetheless, Cleland's widely used computer programs for analysing enzyme kinetic data choose the weighting according to just this sort of reasoning.

1.6   Fitting the straight line

The straight line provides a convenient starting point for discussing the general principles of fitting models to data by the method of least squares, and may be represented as follows:

$$y = \alpha + \beta x + \varepsilon, \tag{1.34}$$

in which α and β are the true values of the parameters to be estimated, y

is the value of the dependent variable observed at a value x of the independent variable, and ε is the error in y, the difference between the observed value y and the true value $\alpha + \beta x$. As eqn 1.34 contains three unknowns, α, β, and ε, but defines only one relation between them, it cannot be solved. Moreover, the problem cannot be overcome by making additional observations of y at further values of x, because each additional observation introduces an additional unknown error ε. No matter how many observations there may be, the number of unknowns always exceeds the number of equations by two.

However, if we replace the true but unhelpful eqn 1.34 by an equation in which α, β, and ε are replaced by estimates a, b, and e,

$$y = a + bx + e, \tag{1.35}$$

then we can try to find estimates that minimize the deviations e of the observed from the calculated y values. It is not normally possible to find parameter estimates that make all e values zero, and in practice, therefore, we must define a function that measures the quality of the fit of any model to the data as a whole. The most obvious function for this purpose might seem to be the simple sum of deviations Σe, but this must be rejected because the e values can be either positive or negative and it is easy to find models that fit the data very badly yet have $\Sigma e = 0$ because the negative and positive deviations cancel exactly. Moreover, such a function would take no account of the possibility discussed in the previous section that the y values are not uniformly reliable.

It is necessary, therefore, to define a weighted function to which all deviations make a positive contribution, either because negative signs are ignored, as in the *sum of absolute deviations SA*,

$$SA = \sum |w^{1/2}e|, \tag{1.36}$$

or because negative values are eliminated by squaring the deviations, as in the *sum of squares of deviations SS*,

$$SS = \sum we^2. \tag{1.37}$$

In these equations the weights w are inversely proportional to the variances of the y values, as defined by eqn 1.27. They are raised to the power $\frac{1}{2}$ in eqn 1.36 so that the same set of w values would be appropriate for minimizing either SA or SS.

There are no strong theoretical objections to defining the best-fit parameter values as those that minimize SA (despite what is claimed in some textbooks); there are, however, practical difficulties in doing so, because the absolute values in eqn 1.36 generate discontinuities in the partial derivatives of SA with respect to the parameter values, so that the methods of differential calculus cannot be used for minimizing SA. Normal practice, therefore, is to define the best-fit values $a = \hat{a}$ and $b = \hat{b}$ as the *least-squares values*, those that make SS a minimum. These may be found by combining eqns 1.35 and 1.37 to show SS as an explicit function of a and b,

$$SS = \sum w(y - a - bx)^2, \tag{1.38}$$

and partially differentiating with respect to a and b:

$$\frac{\partial SS}{\partial a} = -2 \sum w(y - a - bx), \tag{1.39}$$

$$\frac{\partial SS}{\partial b} = -2 \sum wx(y - a - bx). \tag{1.40}$$

Defining $a = \hat{a}$ and $b = \hat{b}$ as the values that make these partial derivatives simultaneously zero, we can set up a pair of simultaneous equations in the parameter values $\hat{a}$ and $\hat{b}$:

$$\hat{a} \sum w + \hat{b} \sum wx = \sum wy, \tag{1.41}$$

$$\hat{a} \sum wx + \hat{b} \sum wx^2 = \sum wxy, \tag{1.42}$$

which have the following solutions:

$$\hat{b} = \frac{\sum w \sum wxy - \sum wx \sum wy}{\sum w \sum wx^2 - \left(\sum wx \right)^2}, \tag{1.43}$$

$$\hat{a} = \frac{\sum wy - \hat{b} \sum wx}{\sum w}. \tag{1.44}$$

These, then, are the least-squares estimates of the slope and intercept of a straight line, and the process of calculating them is often called *linear regression*. The precision with which they are estimated can be assessed by recognizing that the right-hand sides of eqns 1.43 and 1.44 are both linear functions of the y values. In the case of eqn 1.43, dividing all terms by

Σw, and writing the weighted mean of the x values, $\Sigma wx/\Sigma w$, as $\bar{x}$, we have

$$\hat{b} = \frac{\sum wxy - \bar{x}\sum wy}{\sum wx^2 - \bar{x}\sum wx} = \frac{\sum w(x - \bar{x})y}{\sum w(x - \bar{x})x} = \sum ux, \qquad (1.45)$$

where

$$u = \frac{w(x - \bar{x})}{\sum w(x - \bar{x})x}. \qquad (1.46)$$

As u is a function only of the x values it is known exactly, and consequently the variance of $\hat{b}$ follows from eqns 1.8 and 1.10:

$$\sigma^2(\hat{b}) = \sum u^2\sigma^2(y). \qquad (1.47)$$

As each weight w has been defined as inversely proportional to the variance of the corresponding y value, $\sigma^2(y)$ can be written as σ_{exp}^2/w, where σ_{exp}^2 is a constant known as the *experimental variance*. Equation 1.47 can then be written as follows:

$$\sigma^2(\hat{b}) = \sigma_{exp}^2 \sum (u^2/w). \qquad (1.48)$$

Provided that the experimental variance can be estimated (as will be discussed in the next section), this equation can be used as it stands, but it is more convenient to avoid the need to calculate the u values and to derive an expression that contains only the summations that already appear in eqn 1.45. So, substituting eqn 1.46 into eqn 1.48 and rewriting $\bar{x}$ as $\Sigma wx/\Sigma w$, it becomes as follows:

$$\sigma^2(\hat{b}) = \frac{\sigma_{exp}^2 \sum w}{\sum w \sum wx^2 - \left(\sum wx\right)^2}. \qquad (1.49)$$

A similar argument leads to the variance of the intercept $\hat{a}$:

$$\sigma^2(\hat{a}) = \frac{\sigma_{exp}^2 \sum wx^2}{\sum w \sum wx^2 - \left(\sum wx\right)^2}. \qquad (1.50)$$

The meaning of the relationship expressed by eqn 1.44 may be seen more clearly by writing $\Sigma wy/\Sigma w$ and $\Sigma wx/\Sigma w$ as the weighted means $\bar{y}$ and $\bar{x}$ respectively,

$$\hat{a} = \bar{y} - \bar{x}\hat{b}, \qquad (1.51)$$

and the covariance of $\hat{a}$ and $\hat{b}$ can now be written as follows:

$$\operatorname{cov}(\hat{a}, \hat{b}) = \operatorname{cov}\left(\bar{y} - \bar{x}\hat{b}, \hat{b}\right) = \operatorname{cov}\left(\bar{y}, \hat{b}\right) - \bar{x}\operatorname{cov}(\hat{b}, \hat{b}) \quad (1.52)$$

because covariances of uncorrelated quantities are, like variances, additive. In this expression $\operatorname{cov}(\bar{y}, \hat{b}) = 0$, because the covariance of a constant and a variable is zero, and $\operatorname{cov}(\hat{b}, \hat{b}) = \sigma^2(\hat{b})$, as it is the covariance of a variable with itself. So eqn 1.52 may be written as follows:

$$\operatorname{cov}(\hat{a}, \hat{b}) = -\bar{x}\sigma^2(\hat{b}) = \frac{-\sigma_{\exp}^2 \sum wx}{\sum w \sum wx^2 - \left(\sum wx\right)^2}. \quad (1.53)$$

1.7   Degrees of freedom

The experimental variance was introduced in the previous section and used in the definitions of the variances and covariances of the parameters of the straight line, but the question of how it might be estimated from a set of experimental observations was left unanswered. The ideal estimate would be in terms of the sum of squares of true errors, i.e.

$$n\sigma_{\exp}^2 = \sum w(y - \alpha - \beta x)^2, \quad (1.54)$$

but this is not possible because the true parameter values α and β remain unknown. The obvious alternative is to replace $n\sigma_{\exp}^2$ by SS, i.e. the corresponding estimate based on the best-fit estimates of the parameters

$$SS = \sum w\left(y - \hat{a} - \hat{b}x\right)^2. \quad (1.55)$$

The algebra needed to compare $n\sigma_{\exp}^2$ with SS can be greatly simplified by assuming, without loss of generality, that the origin is defined in such a way that Σwx and Σwy are both zero, in which case eqns 1.43 and 1.44 can be written in the following much simpler forms:

$$\hat{b} = \sum wxy \Big/ \sum wx^2, \quad (1.56)$$

$$\hat{a} = 0. \quad (1.57)$$

That this simplification does not imply any loss of generality may be seen by considering that it can be introduced *after* estimating and drawing a best-fit straight line; this implies only that the axes are moved, leaving the points, the best-fit line, and the scatter of points about the line unchanged. Note also that eqn 1.57 does not imply either that $\alpha = 0$ or that $\sigma^2(\hat{a}) = 0$.

If now we expand the right-hand sides of eqns 1.54 and 1.55 and subtract one from the other, we have

$$n\sigma_{\mathrm{exp}}^2 - SS = \alpha^2 \sum w + \left(\hat{b} - \beta\right)^2 \sum wx^2$$

$$= (\hat{a} - \alpha)^2 \sum w + \left(\hat{b} - \beta\right)^2 \sum wx^2. \qquad (1.58)$$

(For converting the middle form of this equation to the right-hand side, recall that we have assumed $\hat{a} = 0$.) The expected values of $(\hat{a} - \alpha)^2$ and $(\hat{b} - \beta)^2$ are the variances of $\hat{a}$ and $\hat{b}$ respectively, which are known from eqns 1.49 and 1.50, and so it follows that the expected value of $n\sigma_{\mathrm{exp}}^2 - SS$ is given by

$$\mathrm{E}\left(n\sigma_{\mathrm{exp}}^2 - SS\right) = \sigma^2(\hat{a}) \sum w + \sigma^2(\hat{b}) \sum wx^2 = 2\sigma_{\mathrm{exp}}^2, \qquad (1.59)$$

i.e.

$$\mathrm{E}(SS) = (n - 2)\sigma_{\mathrm{exp}}^2, \qquad (1.60)$$

and thus the best unbiassed estimate of σ_{exp}^2 for use in eqns 1.49, 1.50, and 1.53 is

$$\widehat{\sigma_{\mathrm{exp}}^2} = SS/(n - 2). \qquad (1.61)$$

In this expression the divisor $n - 2$ is known as the number of *degrees of freedom*; it may be compared with the factor $n - 1$ that appeared in eqn 1.17. In general, the number of degrees of freedom is less than the number of observations, differing from it by the number of parameters that have been estimated; for obtaining unbiassed estimates of the experimental variance and other variances calculated from it, the minimum sum of squares must be divided by the number of degrees of freedom, not by the number of observations.

If the solutions to the weighted least-squares analysis of the straight line, given in eqns 1.43 and 1.44, are compared with the more familiar equations found in many elementary textbooks for unweighted linear regression, it will be seen that the latter can be obtained from eqns 1.43 and 1.44 by replacing Σw by n, the number of observations. This is obviously correct if each $w = 1$, but it is also correct if each $w = c$, where c is any constant, because c will be a factor of every term in every summation and will cancel from the final expressions. Unfortunately, from the biochemist's point of view, it is virtually never realistic to assume that biochemical data are adequately described by a straight line with equal variance for each observation. This provides the most important reason for regarding weighted least-squares analysis as fundamental, and for forgetting about unweighted linear least squares. There is, however, a second reason that is illustrated by the discussion of degrees of freedom in this section: although it is always possible to convert the weighted solution of a least-squares problem into the corresponding unweighted solution by replacing Σw by n and otherwise omitting w wherever it occurs, the transformation cannot be made in reverse, because the weighted solution cannot simply be written down from the unweighted solution by assuming that every n must be replaced by Σw and that w must be included in every other summation. Sometimes, as in consideration of degrees of freedom, n must remain as n.

1.8   Choice of dependent variable

The numerator of the expression for the slope, eqn 1.43, is symmetrical with respect to interchange of the two variables, but the denominator is not, as it is a function only of the independent variable x and does not contain the dependent variable y. This asymmetry follows through to the other equations that were obtained using eqn 1.43, and is a direct consequence of a corresponding asymmetry in the initial assumptions, when we assumed that x was known exactly but that y was subject to error. It follows that treating x as the dependent variable will not yield the same straight line as treating y as the dependent variable and, as illustrated in Fig. 1.3, the difference between the two is not trivial, especially if the scatter of observations is substantial. In the limit when the scatter is so great that there is no correlation between x and y, the two regression lines are at right angles to one another. Thus one ought not to be too casual about deciding which to treat as the independent variable. In enzyme kinetics it is nearly always reasonable to assume that substrate concentrations are known much more accurately than rates, so that it is not too big an assumption to treat all the error as applying to the rate, and this is what is normally done.

If both variables are subject to error and neither can be ignored, the problem becomes much more complicated and will not be discussed in this book, beyond noting that fits based on maximum likelihood (Section 4.3) can be obtained, and emphasizing that simple compromises between the two limiting cases shown in the top two panels of Fig. 1.3 are *not* normally appropriate. The fundamental problem that arises if deviations are measured in any direction other than parallel with one or other axis is that such deviations do not have properly defined dimensions (except in the unusual case where x and y have the same dimensions and fall naturally on the same scale). The practical consequence of this is that the fit obtained in such a case will depend arbitrarily on the scales chosen for plotting. The fit shown in the bottom panel of Fig. 1.3 was obtained by

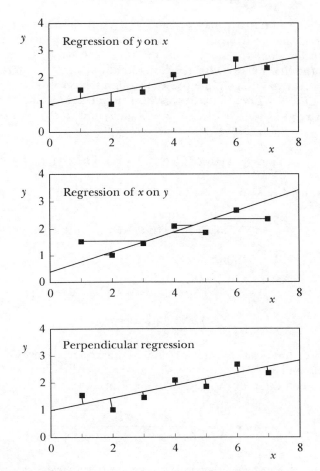

Fig. 1.3. *Three non-equivalent regressions.* Treating y as the dependent variable (*top*) does not give the same result as treating x as the dependent variable (*middle*), and neither corresponds to minimizing the sum of squares of deviations measured along the shortest distance (*bottom*). The set of (x, y) values was as follows: (1, 1.55), (2, 1.02), (3, 1.47), (4, 2.11), (5, 1.87), (6, 2.68), (7, 2.36). The fitted lines were as follows: top, $y = 0.9871 + 0.2196\,x$; middle, $x = -0.9841 + 2.671\,y$, or $y = 0.3684 + 0.3743\,x$; bottom, $y = 0.9581 + 0.2269\,x$.

minimizing the sum of squares of perpendicular deviations from the line. It resembles the regression of y on x much more closely than that of x on y, but this is a simple artefact of the scales and units. Because these are such that the trend of points is much closer to horizontal than to vertical, perpendicular deviations are not very different from vertical deviations.

Although minimizing the sum of squares of perpendicular deviations is not usually appropriate, it may become so if both variables are subject to error and both derive from the same kinds of measurement. In this case it is easiest to consider the slope b in the form $b = \tan \theta$, where θ is the angle that the line makes with the horizontal axis, and to start from an assumption that the best-fit line passes through the centroid $(\bar{x}, \bar{y})$, where $\bar{x} = \Sigma wx / \Sigma w$, $\bar{y} = \Sigma wy / \Sigma w$. In this case the perpendicular deviation of any point (x, y) is $(y - \bar{y})\cos \theta - (x - \bar{x})\sin \theta$, and so the sum of squares SS is

$$SS = \sum w[(y - \bar{y})\cos \theta - (x - \bar{x})\sin \theta]^2. \qquad (1.62)$$

Differentiating this with respect to θ gives

$$\frac{dSS}{d\theta} = 2 \sum w[(y - \bar{y})\cos \theta - (x - \bar{x})\sin \theta]$$

$$\times [-(y - \bar{y})\sin \theta - (x - \bar{x})\cos \theta]$$

$$= \sin 2\theta \left[\sum w(x - \bar{x})^2 - \sum w(y - \bar{y})^2 \right]$$

$$- 2 \cos 2\theta \sum w(x - \bar{x})(y - \bar{y}). \qquad (1.63)$$

(The second line makes use of the standard relationships $\sin 2\theta = 2 \sin \theta \cos \theta$ and $\cos 2\theta = \cos^2 \theta - \sin^2 \theta$.) Setting this derivative to zero then allows definition of the best-fit angle $\hat{\theta}$ in terms of

$$\tan 2\hat{\theta} = \frac{2 \sum w(x - \bar{x})(y - \bar{y})}{\sum w(x - \bar{x})^2 - \sum w(y - \bar{y})^2}, \qquad (1.64)$$

from which calculation of the slope as $\tan \hat{\theta}$ is then straightforward.

2
Fitting the Michaelis–Menten equation by least squares

2.1 Linearization of the Michaelis–Menten equation

The Michaelis–Menten equation, together with virtually all of the other models that are of interest in enzyme kinetics, differs from the straight line equation in the fundamental respect that it is a *non-linear equation*. This is not simply a trivial restatement of the fact that as ordinarily written it does not define a straight line. In the context of statistical analysis linearity refers not to the relationship between the variables but to the relationship between the dependent variable and the unknown parameters; a model is linear if the dependent variable can be expressed as the sum of a set of terms of which each is either constant or proportional to one parameter, i.e. it is of the following form:

$$y_i = \beta_0 + \beta_1 x_{1i} + \beta_2 x_{2i} + \cdots + \varepsilon_i, \tag{2.1}$$

where y_i is the ith of a set of values of the dependent variable y, assumed to be the only variable subject to error, observed at values $x_{1i}, x_{2i} \ldots$ of the independent variables; $b_0, b_1, b_2 \ldots$ are the parameters of the model, and ε_i is the error. The equation for a polynomial dependence of y on x, for example

$$y = \beta_0 + \beta_1 x + \beta_2 x^2 + \beta_3 x^3 + \cdots + \varepsilon, \tag{2.2}$$

is of the form of eqn 2.1 and is a linear equation even though it does not define a straight-line dependence of y on x.

The importance of linearity is that any linear model, no matter how many independent variables and parameters it may contain, can be analysed by generalizing the methods used for the straight line, because it is always possible to set up and solve a set of simultaneous equations similar to eqns 1.41 and 1.42. If the Michaelis–Menten equation is written with an explicit error term, however, it is not of the form of eqn 2.1,

$$v = \frac{Va}{K_m + a} + e, \tag{2.3}$$

and is accordingly non-linear. (Here the error is written as e and not as ε to indicate that it expresses a deviation from a model rather than a true error.) Although one can easily define a sum of squares SS as Σwe^2, and can partially differentiate it to obtain expressions for the partial derivatives with respect to V and K_m, setting these partial derivatives to zero leads to a pair of non-linear simultaneous equations that has no analytical solution. Moreover, although the Michaelis–Menten equation can easily be transformed into a straight-line equation if written without the error term e, this cannot be done when e is included. If an error term is included after transformation, as in

$$\frac{a}{v} = \frac{K_m}{V} + \frac{1}{V}\, a + e_{\text{lin}}, \qquad (2.4)$$

then the result is not a true transformation of eqn 2.3, e_{lin} is not the same as e, and minimizing the linear sum of squares $SS_{\text{lin}} = \Sigma e_{\text{lin}}^2$ does not provide the same parameter values as minimization of the true sum of squares SS. Nonetheless, by examining the relationship between e and e_{lin} it is still possible to use the methods of linear regression to minimize SS:

$$e = \frac{-Vve_{\text{lin}}}{K_m + a}. \qquad (2.5)$$

This can be used to express SS in terms of e_{lin}:

$$SS = \sum we^2 = \sum \frac{wV^2 v^2 e_{\text{lin}}^2}{(K_m + a)^2} = \sum \frac{wv^2 \hat{v}^2 e_{\text{lin}}^2}{a^2} \approx \sum \frac{wv^4 e_{\text{lin}}^2}{a^2}, \qquad (2.6)$$

where $\hat{v} = Va/(K_m + a)$ is the v value calculated from any (not necessarily best-fit) parameter values. (It might seem more logical to reserve the symbol $\hat{v}$ for the v values calculated from the best-fit model, but as it is mainly needed for the actual process of calculating the best-fit model, it is more convenient to give it a more general meaning.)

It now follows from eqn 2.6 that SS can be minimized by minimizing SS_{lin} defined with the weights w appropriate for v replaced by weights $wv^2 \hat{v}^2/a^2$ for a/v. As a first approximation, however, it is necessary to use weights wv^4/a^2 calculated from the observed rates v, because these are known whereas no estimates of V and K_m exist for use in calculating the $\hat{v}$ values. These first-approximation weights provide estimates of V and K_m that can be used for calculating improved weights $wv^2 \hat{v}^2/a^2$, from which improved estimates of V and K_m can be obtained and the process continued until these estimates are self-consistent, i.e. they do not change from one iteration to the next.

Table 2.1 Weights for a/v in fitting the Michaelis–Menten equation

The Table compares the results of fitting data to the Michaelis–Menten equation by non-linear regression and by linear regression of the plot of a/v against a with various approximations to the proper weights. Note that weights of $\hat{v}^3 v/a^2$ for a/v give parameter values identical with those given by non-linear regression with uniform weights for v. The data used for the calculation were used by Wilkinson (1961) to illustrate a non-linear regression method, and consisted of six (a, v) pairs as follows: $(0.138, 0.148)$, $(0.220, 0.171)$, $(0.291, 0.234)$, $(0.560, 0.324)$, $(0.766, 0.390)$, and $(1.46, 0.493)$. The corresponding results for the double-reciprocal plot of $1/v$ against $1/a$ may be obtained by replacing a by 1 throughout the table, e.g. $w(1/v) = v^4$ gives identical parameter estimates, $\hat{K}_m = 0.57097$ and $\hat{V} = 0.67986$, with those shown for $w(a/v) = v^4/a^2$. The table is based on results of Cornish-Bowden (1982).

Weighting	$\hat{K}_m$	$\hat{V}$	Reference
Unweighted non-linear regression			
$w(v) = 1$	0.59655	0.69040	Wilkinson 1961
Weighted linear regression			
$w(a/v) = v^4/a^2$	0.57097	0.67986	Burk 1934; Lineweaver *et al.* 1934; Johansen and Lumry 1961; Wilkinson 1961
$w(a/v) = \hat{v}^2 v^2/a^2$	0.58770	0.68672	Cornish-Bowden 1976
$w(a/v) = \hat{v}^3 v/a^2$	0.59655	0.69040	Cornish-Bowden 1982
$w(a/v) = \hat{v}^4/a^2$	0.60553	0.69416	Cleland 1967
Unweighted linear regression			
$w(a/v) = 1$	0.58209	0.68477	
$w(1/v) = 1$	0.44062	0.58532	

This iterative procedure is essentially as I have suggested in the past (Cornish-Bowden 1976), but it differs from that suggested by Cleland (1967), in which the initial weights wv^4/a^2 would be replaced by refined weights $w\hat{v}^4/a^2$ in the subsequent iterations. The comparison shown in Table 2.1 of the results obtained by the two procedures, however, shows that they both differ from those given by minimizing SS directly by non-linear regression as suggested by Wilkinson (1961). Clearly, therefore, there must be a fault in the argument outlined above. The explanation is that in partially differentiating eqn 1.38 in the process of deriving eqns 1.43 and 1.44, the least-squares solution for the straight line, the weights were treated as constants independent of the parameter values. In the case of the Michaelis–Menten equation, however, the weights depend on the values of V and K_m. It turns out (Cornish-Bowden 1982) that the inaccuracy can be exactly corrected by using refined weights $wv\hat{v}^3/a^2$ rather than $wv^2\hat{v}^2/a^2$ in the iterative process, as may be seen in Table 2.1.

A slight complication is that this is a correction of the logical error of treating weights as constants during the differentiation of the sum of squares; for calculating the value of the sum of squares no differentiation is involved, so no correction is needed. What this means in practice is that once the best fit has been found one should revert to final weights $wv^2\hat{v}^2/a^2$ for calculating the sum of squares and information derived from it, such as parameter variances and covariances. More generally, wherever a factor $\hat{v}^3$ appears in the minimization it should be replaced by $v\hat{v}^2$ at the end.

2.2   Corresponding results for the double-reciprocal plot

I have preferred to give this discussion in relation to the linear plot of a/v against a rather than the *double-reciprocal plot* of $1/v$ against $1/a$, despite the fact that the latter is more familiar to most biochemists and much more widely encountered in textbooks and research papers. This is because v^4/a^2 does not deviate from constancy nearly as grossly as v^4 does, and consequently it matters less in the case of the plot of a/v against a if the proper weighting is simply ignored. To be more explicit, the following equation can be written instead of eqn 2.4,

$$\frac{1}{v} = \frac{1}{V} + \frac{K_m}{V}\frac{1}{a} + e'_{\text{lin}}, \tag{2.7}$$

and essentially the same logic applied to it. This results in the same weights as before without the factors of $1/a^2$, i.e. wv^4/a^2 in the discussion above

Table 2.2   Weighting in different linear transformations

The table compares the weights for $1/v$ with those for a/v for data following the Michaelis–Menten equation $v = Va/(K_m + a)$ assuming either a uniform variance of v, or a uniform coefficient of variation of v. The range of concentrations considered is chosen to correspond to the sort of range often used in real experiments. The weights are calculated from eqns 1.30–1.33, but in each case the constant of proportionality is adjusted to give a weight of unity for a point with $a = K_m$. The bottom line of the table shows the ratio of the largest to the smallest of the values in each column.

| a/K_m | v/V | Uniform variance in v | | Uniform coefficient of variation in v | |
		$w(1/v)$	$w(a/v)$	$w(1/v)$	$w(a/v)$
0.1	0.091	0.00109	0.109	0.0331	3.31
0.2	0.167	0.012	0.309	0.111	2.78
0.5	0.333	0.198	0.790	0.444	1.78
1.0	0.500	1.00	1.00	1.00	1.00
2.0	0.667	3.16	0.790	1.78	0.444
5.0	0.833	7.71	0.309	2.78	0.111
Range	9.17	7073	9.17	84.0	29.8

must be replaced by $w v^4$, $w \hat{v}^2 v^2 / a^2$ by $w \hat{v}^2 v^2$, $w \hat{v}^4 / a^2$ by $w \hat{v}^4$, and $w \hat{v}^3 v / a^2$ by $w \hat{v}^3 v$. The variations in weights for $1/v$ are compared in Table 2.2 with those for a/v, for two different assumptions about the weights w that apply to v (see next section). Notice that in one case the weights for $1/v$ vary over a factor of more than 7000 whereas those for a/v vary by less than a factor of 10; the other case is less striking, but even then the weights for $1/v$ vary more than those for a/v.

2.3 Choosing the proper weights for the rate

Before the results of the previous sections can be applied in practice, an appropriate weighting scheme for v is needed, so that values can be assigned to the factors w that appear in all the summations. It has sometimes been recommended that in the absence of other information each v should be assumed to have the same variance, a case that I shall refer to as *simple errors*, but most of the experimental investigations that have been made of error behavior in enzyme kinetics (e.g. Siano *et al.* 1975; Storer *et al.* 1975; Askelöf *et al.* 1976) have suggested that the truth is more often closer to a constant coefficient of variation for v, or *relative errors*. (Discussion of more complex possibilities such as behaviour intermediate between simple and relative errors will be deferred until Section 6.3.)

For simple errors, the weights for v are simply $w = 1$ for all observations. These imply preliminary weights of v^4 / a^2 for a/v or (equivalently) v^4 for $1/v$. In either case, therefore, substitution into eqns 1.43 and 1.44 gives, as preliminary estimates,

$$\frac{1}{\hat{V}} = \frac{\left(\sum v^4 / a^2 \right) \sum v^3 - \left(\sum v^4 / a \right) \left(\sum v^3 / a \right)}{\left(\sum v^4 / a^2 \right) \sum v^4 - \left(\sum v^4 / a \right)^2}, \qquad (2.8)$$

$$\frac{\hat{K}_m}{\hat{V}} = \frac{\sum v^4 \left(\sum v^3 / a \right) - \left(\sum v^4 / a \right) \sum v^3}{\left(\sum v^4 / a^2 \right) \sum v^4 - \left(\sum v^4 / a \right)^2}. \qquad (2.9)$$

These preliminary parameter values may now be used to calculate revised weights $\hat{v}^3 v / a^2$ for a/v (or $\hat{v}^3 v$ for $1/v$), and hence refined parameter estimates as follows:

$$\frac{1}{\hat{V}} = \frac{\left(\sum \hat{v}^3 v / a^2 \right) \sum \hat{v}^3 - \left(\sum \hat{v}^3 v / a \right) \left(\sum \hat{v}^3 / a \right)}{\left(\sum \hat{v}^3 v / a^2 \right) \sum \hat{v}^3 v - \left(\sum \hat{v}^3 v / a \right)^2}, \qquad (2.10)$$

$$\frac{\hat{K}_m}{\hat{V}} = \frac{\sum \hat{v}^3 v \left(\sum \hat{v}^3 / a \right) - \left(\sum \hat{v}^3 v / a \right) \sum \hat{v}^3}{\left(\sum \hat{v}^3 v / a^2 \right) \sum \hat{v}^3 v - \left(\sum \hat{v}^3 v / a \right)^2}. \qquad (2.11)$$

These last equations may then be used iteratively until the values do not change from one iteration to the next. This typically requires about four iterations.

Once the best-fit values of $1/V$ and K_m/V are known they may be used to calculate best-fit values of V, K_m, etc. Strictly speaking, although these secondary parameters are indeed least-squares estimates of the parameters in question, they are not unbiassed estimates, i.e. for infinite sample sizes they do not converge to the population means. This is because, as mentioned in Section 1.4, a non-linear function of an unbiassed estimate of a parameter is not an unbiassed estimate of the same function of the same

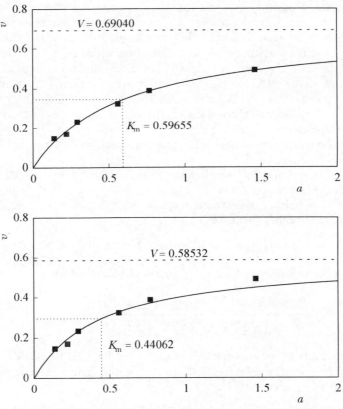

Fig. 2.1. *Two 'best-fit' curves.* Both plots show the data of Wilkinson (1961), listed in the caption to Table 2.1. The upper curve is a best-fit line calculated by unweighted non-linear regression; the lower is a transformation to (a, v) coordinates of the results of unweighted linear regression of the double-reciprocal plot. Note especially the point at $a = 1.46$, which shows that the appearance of a good fit in a double-reciprocal plot does not imply that the original measurements are well fitted.

parameter. The bias is usually small, however, and is almost always ignored in practical applications of non-linear regression. Moreover, pedantically accurate symbols like $1/\widehat{(1/V)}$ rather than simply $\hat{V}$ are awkward both to write and to print (with the circumflex over the whole parameter) and unnecessarily cumbersome.

Examination of a numerical example, such as the one shown in Table 2.1, shows that the advantage of using eqns 2.10 and 2.11 to obtain refined parameter estimates (or of using the direct non-linear regression method of Wilkinson (1961), which is no less laborious than the procedure described here) is often very slight, because the final values (lines 1 or 4 of Table 2.1) differ from the preliminary values (line 2) by amounts that are small compared with the statistical errors in the results. This is true even if the assumption of simple errors is known to be exactly correct, but in reality it may not be and it is arguable whether it is worthwhile to obtain an exact solution to a somewhat inexact question. Nonetheless, with modern computing equipment the additional effort is almost trivial.

One should contrast the rather slight advantage in using refined weights, which is not a peculiarity of the data set considered in Table 2.1 but is a common experience, with the substantial advantage in using weights at all: the bottom line of the table shows that the unweighted estimate of K_m given by linear regression of the plot of $1/v$ against $1/a$ differs from the true least-squares estimate by more than 25 per cent and if one transforms the results from the linear regression back to (a, v) coordinates (Fig. 2.1), their inadequacy becomes obvious to the eye. The unweighted regression of the plot of a/v against a (line 6 of Table 2.1) performs much better, as expected from the comparison shown in Table 2.2. It is, however, a chance characteristic of the data set that in this example this plot gives almost perfect results without weighting.

When the errors are relative, i.e. when each rate can be assumed to have a uniform coefficient of variation, the analysis is somewhat simpler because the preliminary fit provides an exact least-squares solution without any need for iterative refinement of the weights. The required weights for v are now $1/\hat{v}^2$, which means that $w\hat{v}^2 = 1$ and thus cancels from the exact form of eqn 2.6. Consequently the preliminary weights for a/v and $1/v$ are also exact weights and we can write the following expressions for estimating V and K_m in a single step:

$$\frac{1}{\hat{V}} = \frac{\left(\sum v^2/a^2\right)\sum v - \left(\sum v^2/a\right)\left(\sum v/a\right)}{\left(\sum v^2/a^2\right)\sum v^2 - \left(\sum v^2/a\right)^2}, \tag{2.12}$$

$$\frac{\hat{K}_m}{\hat{V}} = \frac{\sum v^2\left(\sum v/a\right) - \left(\sum v^2/a\right)\sum v}{\left(\sum v^2/a^2\right)\sum v^2 - \left(\sum v^2/a\right)^2}. \tag{2.13}$$

2.4   Standard errors of Michaelis–Menten parameters

The expression derived in Section 1.6 for the variances of the parameters of the straight line, i.e. eqns 1.49, 1.50, and 1.53, can be applied directly to $1/V$ and K_m/V, as long as it is remembered that the required weights are those for the straight-line transformation of the Michaelis–Menten equation, not the weights that apply to v itself. Making the appropriate substitutions, therefore, the case of simple errors gives the following expressions for the variances:

$$\sigma^2(1/\hat{V}) = \frac{\sigma^2_{\exp} \sum v^2 \hat{v}^2/a^2}{\left(\sum v^2 \hat{v}^2/a^2\right) \sum v^2 \hat{v}^2 - \left(\sum v^2 \hat{v}^2/a\right)^2}, \quad (2.14)$$

$$\sigma^2(\hat{K}_m/\hat{V}) = \frac{\sigma^2_{\exp} \sum v^2 \hat{v}^2}{\left(\sum v^2 \hat{v}^2/a^2\right) \sum v^2 \hat{v}^2 - \left(\sum v^2 \hat{v}^2/a\right)^2}, \quad (2.15)$$

$$\mathrm{cov}(1/\hat{V}, \hat{K}_m/\hat{V}) = \frac{\sigma^2_{\exp} \sum v^2 \hat{v}^2/a}{\left(\sum v^2 \hat{v}^2/a^2\right) \sum v^2 \hat{v}^2 - \left(\sum v^2 \hat{v}^2/a\right)^2}. \quad (2.16)$$

The variances of additional parameters such as V and K_m can be obtained by applying the results of Section 1.4 to these equations. For example, the variance of V follows simply by combining eqn 1.23 with eqn 2.14:

$$\sigma^2(\hat{V}) \approx \frac{\hat{V}^2 \sigma^2_{\exp} \sum v^2 \hat{v}^2/a^2}{\left(\sum v^2 \hat{v}^2/a^2\right) \sum v^2 \hat{v}^2 - \left(\sum v^2 \hat{v}^2/a\right)^2}. \quad (2.17)$$

The variance of K_m follows by applying eqn 1.20 to eqns 2.14–2.16 (treating K_m as $(K_m/V)/(1/V)$):

$$\sigma^2(\hat{K}_m) \approx \frac{\hat{V}^2 \sigma^2_{\exp}\left(\sum v^2 \hat{v}^2 + 2\hat{K}_m \sum v^2 \hat{v}^2/a + \hat{K}_m^2 \sum v^2 \hat{v}^2/a^2\right)}{\left(\sum v^2 \hat{v}^2/a^2\right) \sum v^2 \hat{v}^2 - \left(\sum v^2 \hat{v}^2/a\right)^2}$$

$$= \frac{\hat{V}^2 \sigma^2_{\exp} \sum \left[(1 + \hat{K}_m/a)^2 v^2 \hat{v}^2\right]}{\left(\sum v^2 \hat{v}^2/a^2\right) \sum v^2 \hat{v}^2 - \left(\sum v^2 \hat{v}^2/a\right)^2}. \quad (2.18)$$

The covariance of V and K_m is more difficult to derive, though the result is what one would guess by inspecting eqns 2.17 and 2.18 and 'interpolating'. To derive it without guessing, we use eqn 1.20 to write

$$\sigma^2\left(\frac{\hat{K}_m}{\hat{V}}\right) \approx \frac{\sigma^2(\hat{K}_m)}{\hat{V}^2} - \frac{2\hat{K}_m\,\mathrm{cov}\left(\hat{K}_m, \hat{V}\right)}{\hat{V}^3} + \frac{\hat{K}_m^2\sigma^2(\hat{V})}{\hat{V}^4}, \quad (2.19)$$

and then substitute the first, second, and fourth terms by means of eqns 2.15, 2.17, and 2.18, and rearrange:

$$\mathrm{cov}\left(\hat{K}_m, \hat{V}\right) \approx \frac{\hat{V}^2\sigma_{\mathrm{exp}}^2\left(\hat{K}_m\sum v^2\hat{v}^2/a^2 + \sum v^2\hat{v}^2/a\right)}{\left(\sum v^2\hat{v}^2/a^2\right)\sum v^2\hat{v}^2 - \left(\sum v^2\hat{v}^2/a\right)^2}$$

$$= \frac{\hat{V}^2\sigma_{\mathrm{exp}}^2\sum\left[\left(1 + \hat{K}_m/a\right)v^2\hat{v}^2/a\right]}{\left(\sum v^2\hat{v}^2/a^2\right)\sum v^2\hat{v}^2 - \left(\sum v^2\hat{v}^2/a\right)^2}. \quad (2.20)$$

In all cases the *experimental variance* σ_{exp}^2 may be estimated as $SS/(n-2)$, and the *standard error* of each parameter estimate may be taken as the square root of the corresponding variance estimate. The covariances defined by eqns 2.16 and 2.20 may appear to be of less importance, but in fact they provide valuable information about the nature of the best-fit parameter values, as we shall see. They are most easily interpreted by converting them to the dimensionless form expressed by *Pearson's correlation coefficient* (also known as the *product–moment correlation coefficient* or often just the *correlation coefficient*), which is defined for any pair of variables x and y as

$$\rho(x, y) = \frac{\mathrm{cov}(x, y)}{\sigma(x)\sigma(y)}. \quad (2.21)$$

A correlation coefficient of ± 1 means that there is perfect linear correlation between x and y, so that the deviation of any y value from its population mean is exactly proportional to the corresponding deviation in x, with a positive constant of proportionality if the correlation coefficient is $+1$ or negative if it is -1. A value of 0 means that there is no linear correlation between x and y.

Estimates of the parameters of the Michaelis–Menten equation are often highly correlated, as may be illustrated by considering again the data listed in Table 2.1, for which values of the standard errors and correlation coefficients are given in Table 2.3. Note especially the coefficient of $+0.94$ for correlation between the estimates of V and K_m: values as high as this are quite typical, and values greater than $+0.99$ are not uncommon for

Table 2.3   Standard errors and correlation coefficients

The table shows the standard errors and correlation coefficients obtained by applying the equations given in Section 2.6 to the data of Table 2.1, assuming simple errors (as in Table 2.1).

Parameter	Value	Standard error	Equations used
$1/\hat{V}$	1.448	0.077	2.12, 2.14
$\hat{K}_m/\hat{V}$	0.864	0.057	2.13, 2.15
$\rho(1/\hat{V}, \hat{K}_m/\hat{V})$	-0.82		2.16, 2.21
$\hat{V}$	0.690	0.036	2.12, 2.17
$\hat{K}_m$	0.597	0.068	2.12, 2.13, 2.18
$\rho(\hat{V}, \hat{K}_m)$	$+0.94$		2.20, 2.21

poorly designed experiments with a range of substrate concentrations that is low in relation to K_m. They mean that almost all of the effect of changing the estimate of V can be compensated for, with very little effect on the sum of squares, by changing the estimate of K_m by an amount sufficient to keep the estimate of V/K_m constant.

All of the equations given in this section refer to the case of simple errors. However, the corresponding equations for relative errors may be obtained very simply by replacing $v^2\hat{v}^2$ by v^2 throughout; for example, applying this substitution to eqn 2.14 gives

$$\sigma^2(1/\hat{V}) = \frac{\sigma_{\text{exp}}^2 \sum v^2/a^2}{\left(\sum v^2/a^2\right) \sum v^2 - \left(\sum v^2/a\right)^2}. \tag{2.22}$$

It is useful to note that all of the summations required for evaluating the expressions given in this section are already used for calculating the parameter estimates. Moreover, each expression (for a given weighting system) has the same denominator.

3
More than two parameters

3.1 The general linear model

Although many of the equations of interest in enzyme kinetics contain more than two parameters, most can be regarded as generalized forms of the Michaelis–Menten equation, and can be fitted by similar methods. Before considering these, however, it will be useful to examine a generalized straight line, known as the *general linear model*. In principle, this can contain any number of parameters, but in the belief that it is easier to follow algebra in relation to a specific case, I shall discuss it here in relation to a four-parameter model. This is sufficiently general to introduce some new ideas not covered in the discussion of the straight line, while at the same time just simple enough for all terms to be written explicitly. The four-parameter linear model may be written as follows:

$$y_i = \beta_1 x_{1i} + \beta_2 x_{2i} + \beta_3 x_{3i} + \beta_4 x_{4i} + \varepsilon_{i,\text{lin}}$$

$$= \sum \beta_j x_{ji} + \varepsilon_{i,\text{lin}} \qquad i = 1 \text{ to } n, \ j = 1 \text{ to } 4. \qquad (3.1)$$

Although the subscript 'lin' (for 'linear'), would not usually be considered necessary for presenting the general linear model, it is retained here to preserve the distinctions made in Section 2.1 and, more important, to avoid confusion when we come to apply the general linear model in Section 3.3. In matrix symbolism, eqn 3.1 may be written as follows:

$$\mathbf{y} = \mathbf{x}\boldsymbol{\beta} + \boldsymbol{\varepsilon}_{\text{lin}}, \qquad (3.2)$$

which is a short-hand representation of

$$
\begin{bmatrix} y_1 \\ y_2 \\ y_3 \\ \vdots \\ y_n \end{bmatrix} = \begin{bmatrix} x_{11} & x_{21} & x_{31} & x_{41} \\ x_{12} & x_{22} & x_{32} & x_{42} \\ x_{13} & x_{23} & x_{33} & x_{43} \\ \vdots & \vdots & \vdots & \vdots \\ x_{1n} & x_{2n} & x_{3n} & x_{4n} \end{bmatrix} \begin{bmatrix} \beta_1 \\ \beta_2 \\ \beta_3 \\ \beta_4 \end{bmatrix} + \begin{bmatrix} \varepsilon_{1,\text{lin}} \\ \varepsilon_{2,\text{lin}} \\ \varepsilon_{3,\text{lin}} \\ \vdots \\ \varepsilon_{n,\text{lin}} \end{bmatrix}. \qquad (3.3)
$$

Note that eqn 3.3 is exactly equivalent to eqn 3.1: it is just a different way of saying the same thing. Equation 3.2 is also equivalent, as long as we take the number of parameters to be defined, but it is actually more general, because it applies to any number of parameters, not just to 4. Matrix algebra was developed as a way of solving whole classes of problems at once, problems that would otherwise need to be dealt with one at a time.*

For purposes of illustration we shall consider the *unweighted* sum of squares SS, which is defined for estimates b_j of the parameters β_j as follows:

$$SS = \sum (y_i - b_1 x_{1i} - b_2 x_{2i} - b_3 x_{3i} - b_4 x_{4i})^2$$

$$= e_{lin}^T e_{lin}$$

$$= (y - xb)^T (y - xb), \tag{3.4}$$

where the superscript T defines the *transpose* of a matrix, i.e. the matrix written with the rows written as columns and the columns as rows. Matrices other than square matrices cannot be multiplied by themselves, but they can always be multiplied by their transposes, and premultiplication of a matrix by its transpose is the operation in matrix algebra that corresponds to taking a square in ordinary algebra.

The first partial derivatives of SS with respect to the parameter estimates are as follows:

$$\frac{\partial SS}{\partial b_1} = 2 b_1 \sum x_{1i}^2 + 2 b_2 \sum x_{1i} x_{2i} + 2 b_3 \sum x_{1i} x_{3i}$$

$$+ 2 b_4 \sum x_{1i} x_{4i} - 2 \sum x_{1i} y_i,$$

$$\frac{\partial SS}{\partial b_2} = 2 b_1 \sum x_{1i} x_{2i} + 2 b_2 \sum x_{2i}^2 + 2 b_3 \sum x_{2i} x_{3i}$$

$$+ 2 b_4 \sum x_{2i} x_{4i} - 2 \sum x_{2i} y_i,$$

$$\frac{\partial SS}{\partial b_3} = 2 b_1 \sum x_{1i} x_{3i} + 2 b_2 \sum x_{2i} x_{3i} + 2 b_3 \sum x_{3i}^2$$

$$+ 2 b_4 \sum x_{3i} x_{4i} - 2 \sum x_{3i} y_i,$$

---

* I shall not attempt to give here a formal account of matrix algebra, for which many textbooks exist. Instead I shall show the equations not only in a familiar form with explicit summations but also as matrix equations, and will add comments to the text to facilitate comparison of the matrix expressions with what may be found in such textbooks.

$$\frac{\partial SS}{\partial b_4} = 2b_1 \sum x_{1i} x_{4i} + 2b_2 \sum x_{2i} x_{4i} + 2b_3 \sum x_{3i} x_{4i}$$

$$+ 2b_4 \sum x_{4i}^2 - 2 \sum x_{4i} y_i, \tag{3.5}$$

or

$$\frac{\partial SS}{\partial \mathbf{b}} = 2\mathbf{x}^T\mathbf{x}\mathbf{b} - 2\mathbf{x}^T\mathbf{y}. \tag{3.6}$$

When SS is a minimum, all of these derivatives are zero, and thus

$$\hat{b}_1 \sum x_{1i}^2 + \hat{b}_2 \sum x_{1i} x_{2i} + \hat{b}_3 \sum x_{1i} x_{3i} + \hat{b}_4 \sum x_{1i} x_{4i} = \sum x_{1i} y_i,$$

$$\hat{b}_1 \sum x_{1i} x_{2i} + \hat{b}_2 \sum x_{2i}^2 + \hat{b}_3 \sum x_{2i} x_{3i} + \hat{b}_4 \sum x_{2i} x_{4i} = \sum x_{2i} y_i,$$

$$\hat{b}_1 \sum x_{1i} x_{3i} + \hat{b}_2 \sum x_{2i} x_{3i} + \hat{b}_3 \sum x_{3i}^2 + \hat{b}_4 \sum x_{3i} x_{4i} = \sum x_{3i} y_i,$$

$$\hat{b}_1 \sum x_{1i} x_{4i} + \hat{b}_2 \sum x_{2i} x_{4i} + \hat{b}_3 \sum x_{3i} x_{4i} + \hat{b}_4 \sum x_{4i}^2 = \sum x_{4i} y_i, \tag{3.7}$$

or

$$\mathbf{x}^T\mathbf{x}\hat{\mathbf{b}} = \mathbf{x}^T\mathbf{y}. \tag{3.8}$$

The least-squares estimates $\hat{b}_j$ may now be found by solving the four simultaneous equations in eqn 3.7, and the solution may be written in matrix terms as follows:

$$\hat{\mathbf{b}} = (\mathbf{x}^T\mathbf{x})^{-1}\mathbf{x}^T\mathbf{y}, \tag{3.9}$$

where $(\mathbf{x}^T\mathbf{x})^{-1}$ is the *inverse* or $\mathbf{x}^T\mathbf{x}$. This means that $(\mathbf{x}^T\mathbf{x})^{-1}\mathbf{x}^T\mathbf{x} = \mathbf{I}_4$, where $\mathbf{I}_4$ is the *identity matrix* of order 4, i.e. the 4×4 matrix with each diagonal element equal to 1 and all non-diagonal elements equal to 0.

Note that if eqns 3.8 and 3.9 were ordinary algebraic equations one would follow from the other by the ordinary rules: matrix algebra is designed as far as possible so that its equations behave like simple equations, though they express much more general relationships.

3.2 Standard errors in the general linear model

If calculation of the best-fit parameter estimates $\hat{b}_j$ were the only objective there would be little point in introducing matrix notation: eqn 3.8 may be much more compact and general than eqns 3.7, but in any one problem it is no less laborious to solve, and for readers with little experience in matrix algebra it is much less obvious what it means. Indeed, the process of matrix inversion required for calculating $(\mathbf{x}^T\mathbf{x})^{-1}$ is very similar to, and just as laborious as, the arithmetic needed for solving eqns 3.7 by elementary methods. The advantage of matrix algebra comes when we need to consider results in addition to the best-fit parameter estimates, most particularly their standard errors. If we write $\mathbf{A} = (\mathbf{x}^T\mathbf{x})^{-1}$, and represent the element in the jth row and kth column of $\mathbf{A}$ as A_{kj}, then the definition of an inverse requires that

$$\mathbf{A}\mathbf{A}^{-1} = \mathbf{A}^{-1}\mathbf{A} = \mathbf{I}_4, \tag{3.10}$$

or, considering the jth row and kth column:

$$[\, A_{1j}\, A_{2j}\, A_{3j}\, A_{4j}\,]\begin{bmatrix} \sum x_{1i}x_{ki} \\ \sum x_{2i}x_{ki} \\ \sum x_{3i}x_{ki} \\ \sum x_{4i}x_{ki} \end{bmatrix} = \delta_{jk},$$

i.e.

$$A_{1j}\sum x_{1i}x_{ki} + A_{2j}\sum x_{2i}x_{ki} + A_{3j}\sum x_{3i}x_{ki} + A_{4j}\sum x_{4i}x_{ki}$$

$$= \delta_{jk} \equiv \begin{cases} 1 \text{ for } j = k \\ 0 \text{ for } j \neq k \end{cases}, \tag{3.11}$$

where δ_{jk}, the *Kronecker delta*, is defined by the identity at the right, i.e. it has a value of 1 if $j = k$ and of 0 if $j \neq k$.

The least-squares estimates may be written in terms of $\mathbf{A}$ as

$$\hat{b}_j = [\, A_{1j}\, A_{2j}\, A_{3j}\, A_{4j}\,]\begin{bmatrix} \sum x_{1i}y_i \\ \sum x_{2i}y_i \\ \sum x_{3i}y_i \\ \sum x_{4i}y_i \end{bmatrix} = \sum g_{ji}y_i, \tag{3.12}$$

where

$$g_{ji} = [\, A_{1j}\, A_{2j}\, A_{3j}\, A_{4j}\,]\begin{bmatrix} x_{1i} \\ x_{2i} \\ x_{3i} \\ x_{4i} \end{bmatrix}. \tag{3.13}$$

Now, if the y_i are independently distributed about their true values with uniform variance σ_{exp}^2, each $\hat{b}_j$ is the sum of n independent variates $g_{ji}y_i$, each of which has variance $g_{ji}^2 \sigma_{\exp}^2$. So, from eqn 1.8,

$$\sigma^2(\hat{b}_j) = \sigma_{\exp}^2 \sum g_{ji}^2, \qquad (3.14)$$

or more generally,

$$\text{cov}(\hat{b}_j, \hat{b}_k) = \sigma_{\exp}^2 \sum g_{ji} g_{ki}. \qquad (3.15)$$

The meaning of these two equations is most easily seen by taking a specific example, e.g. $j = 3$, $k = 4$:

$$\text{cov}(\hat{b}_3, \hat{b}_4) = \sigma_{\exp}^2 \sum g_{3i} g_{4i}$$

$$= \sigma_{\exp}^2 \sum \{ (A_{13} x_{1i} + A_{23} x_{2i} + A_{33} x_{3i} + A_{43} x_{4i})$$

$$\times (A_{14} x_{1i} + A_{24} x_{2i} + A_{34} x_{3i} + A_{44} x_{4i}) \}$$

$$= \sigma_{\exp}^2 \left\{ A_{13}[\, A_{14}\ A_{24}\ A_{34}\ A_{44}\,] \begin{bmatrix} \sum x_{1i}^2 \\ \sum x_{1i} x_{2i} \\ \sum x_{1i} x_{3i} \\ \sum x_{1i} x_{4i} \end{bmatrix} \right.$$

$$+ A_{23}[\, A_{14}\ A_{24}\ A_{34}\ A_{44}\,] \begin{bmatrix} \sum x_{1i} x_{2i} \\ \sum x_{2i}^2 \\ \sum x_{2i} x_{3i} \\ \sum x_{2i} x_{4i} \end{bmatrix}$$

$$+ A_{33}[\, A_{14}\ A_{24}\ A_{34}\ A_{44}\,] \begin{bmatrix} \sum x_{1i} x_{3i} \\ \sum x_{2i} x_{3i} \\ \sum x_{3i}^2 \\ \sum x_{3i} x_{4i} \end{bmatrix}$$

$$\left. + A_{34}[\, A_{14}\ A_{24}\ A_{34}\ A_{44}\,] \begin{bmatrix} \sum x_{1i} x_{4i} \\ \sum x_{2i} x_{4i} \\ \sum x_{3i} x_{4i} \\ \sum x_{4i}^2 \end{bmatrix} \right\}. \qquad (3.16)$$

However, **A** is, by definition, the inverse of $\mathbf{x}^{\mathrm{T}}\mathbf{x}$ and so the first three lines of this expression are zero by virtue of eqn 3.11, whereas the last line is A_{34} by virtue of eqn 3.10. Consequently the whole expression may be written much more simply as

$$\mathrm{cov}\left(\hat{b}_3, \hat{b}_4\right) = A_{34}\,\sigma_{\mathrm{exp}}^2, \tag{3.17}$$

or more generally as

$$\left. \begin{aligned} \mathrm{cov}\left(\hat{b}_j, \hat{b}_k\right) &= A_{jk}\,\sigma_{\mathrm{exp}}^2 \\ \sigma^2\left(\hat{b}_j\right) &= A_{jj}\sigma_{\mathrm{exp}}^2 \end{aligned} \right\}. \tag{3.18}$$

This equation now expresses the major result of this section, which is fundamental in calculating the variances and covariances of the parameter estimates in linear models. It may be stated in words as follows: *the variance–covariance matrix for the parameters of a linear model is equal to the product of the experimental variance and the solution matrix.* In matrix symbols it can be written as

$$\mathrm{cov}(\mathbf{b}) = \sigma_{\mathrm{exp}}^2(\mathbf{x}^{\mathrm{T}}\mathbf{x})^{-1}. \tag{3.19}$$

3.3   Application to enzyme inhibition and other kinetic examples

To make the analysis of the general linear model less abstract, this section will show how it can be applied to a specific case, the equation for mixed inhibition, and will then show how nearly all of the equations commonly encountered in enzyme kinetics can be dealt with in the same way.

If an enzyme-catalysed reaction is studied in the presence of a mixed inhibitor I that affects the apparent values of both V and V/K_{m}, then the equation that describes the kinetics is a modified form of eqn 2.3:

$$v = \frac{Va}{K_{\mathrm{m}}(1 + i/K_{\mathrm{ic}}) + a(1 + i/K_{\mathrm{iu}})} + e \tag{3.20}$$

in which i is the concentration of I,* and K_{ic} and K_{iu} are the competitive and uncompetitive inhibition constants respectively. At first sight this is quite different from the equation for the general linear model for four parameters, eqn 3.1, but just as eqn 2.3 could be written as eqn 2.4, a

---

* This should not be confused with the use of i as an index for the ith observation. As eqns 3.21 and 3.22 provide the only context in this book where both are needed together, and as each is quite natural and obvious in its own context, confusion is sufficiently unlikely to justify refraining from introducing an unnatural alternative for one or the other.

linear equation, by writing the equation for a/v, so eqn 3.20 can be rewritten as follows to make the correspondence with eqn 3.1 clearer:

$$\frac{a}{v} = \frac{K_m}{V} + \frac{1}{V}a + \frac{K_m}{VK_{ic}}i + \frac{1}{VK_{iu}}ai + e_{lin} \qquad (3.21)$$
$$\updownarrow \qquad \updownarrow \qquad \updownarrow \qquad \updownarrow \qquad \updownarrow \qquad \updownarrow$$
$$y_i = b_1 x_{1i} + b_2 x_{2i} + b_3 x_{3i} + b_4 x_{4i} + e_{lin}.$$

Equation 3.1 is reproduced here, with the true parameters β_j replaced by estimates b_j and the error term ε_{lin} written as an estimate e_{lin}. The following substitutions are evident from the comparison between the two:

$$y_i = a/v, \quad x_{1i} = 1, \quad x_{2i} = a, \quad x_{3i} = i, \quad x_{4i} = ai. \qquad (3.22)$$

However, before these substitutions are made in eqns 3.7 it is important to remember that e_{lin} is not the same as e, and so the summations in eqns 3.7 must be weighted to allow for this. Conveniently, the relationship between e_{lin} and e is the same as in the comparison between eqns 2.3 and 2.4, and so we can follow the same logic. First we define w as the appropriate weight for v, and then we replace it by $w\hat{v}^3/a^2$ to compensate for the fact that a/v is the dependent variable rather than v. With the introduction of weights and all the substitutions, eqns 3.7 become as follows:

$$\hat{b}_1 \sum w v \hat{v}^3/a^2 + \hat{b}_2 \sum w v \hat{v}^3/a + \hat{b}_3 \sum w v \hat{v}^3 i/a^2$$
$$+ \hat{b}_4 \sum w v \hat{v}^3 i/a = \sum w \hat{v}^3/a$$

$$\hat{b}_1 \sum w v \hat{v}^3/a + \hat{b}_2 \sum w v \hat{v}^3 + \hat{b}_3 \sum w v \hat{v}^3 i/a + \hat{b}_4 \sum w v \hat{v}^3 i = \sum w \hat{v}^3$$

$$\hat{b}_1 \sum w v \hat{v}^3 i/a^2 + \hat{b}_2 \sum w v \hat{v}^3 i/a + \hat{b}_3 \sum w v \hat{v}^3 i^2/a^2$$
$$+ \hat{b}_4 \sum w v \hat{v}^3 i^2/a = \sum w \hat{v}^3 i/a$$

$$\hat{b}_1 \sum w v \hat{v}^3 i/a + \hat{b}_2 \sum w v \hat{v}^3 i + \hat{b}_3 \sum w v \hat{v}^3 i^2/a$$
$$+ \hat{b}_4 \sum w v \hat{v}^3 i^2 = \sum w \hat{v}^3 i. \qquad (3.23)$$

Initially there are no calculated rates $\hat{v}$, so we use observed rates v instead in the first iteration but replace them in later iterations by values $\hat{v}$ calculated from the parameter values of the latest iteration. In each iteration eqns 3.23 thus constitute a set of linear equations that can be solved for the $\hat{b}_j$, and the process is repeated until the solutions are self-consistent, i.e. they remain unchanged from one iteration to the next. We can then make back-substitutions to obtain the parameter values of interest:

$$\hat{K}_{m} = \hat{b}_{1}/\hat{b}_{2}, \quad \hat{V} = 1/\hat{b}_{2}, \quad \hat{K}_{ic} = \hat{b}_{1}/\hat{b}_{3}, \quad \hat{K}_{iu} = \hat{b}_{2}/\hat{b}_{4}. \quad (3.24)$$

Substitution of symbols and expansion of eqn 3.19 then provides the variance–covariance matrix for the linear parameters:

$$
\begin{bmatrix}
\sigma^{2}(\hat{b}_{1}) & \mathrm{cov}(\hat{b}_{1},\hat{b}_{2}) & \mathrm{cov}(\hat{b}_{1},\hat{b}_{3}) & \mathrm{cov}(\hat{b}_{1},\hat{b}_{4}) \\
\mathrm{cov}(\hat{b}_{2},\hat{b}_{1}) & \sigma^{2}(\hat{b}_{2}) & \mathrm{cov}(\hat{b}_{2},\hat{b}_{3}) & \mathrm{cov}(\hat{b}_{2},\hat{b}_{4}) \\
\mathrm{cov}(\hat{b}_{3},\hat{b}_{1}) & \mathrm{cov}(\hat{b}_{3},\hat{b}_{2}) & \sigma^{2}(\hat{b}_{3}) & \mathrm{cov}(\hat{b}_{3},\hat{b}_{4}) \\
\mathrm{cov}(\hat{b}_{4},\hat{b}_{1}) & \mathrm{cov}(\hat{b}_{4},\hat{b}_{2}) & \mathrm{cov}(\hat{b}_{4},\hat{b}_{3}) & \sigma^{2}(\hat{b}_{4})
\end{bmatrix}
$$

$$
= \sigma_{\mathrm{exp}}^{2}
\begin{bmatrix}
\sum wv\hat{v}^{3}/a^{2} & \sum wv\hat{v}^{3}/a & \sum wv\hat{v}^{3}i/a^{2} & \sum wv\hat{v}^{3}i/a \\
\sum wv\hat{v}^{3}/a & \sum wv\hat{v}^{3} & \sum wv\hat{v}^{3}i/a & \sum wv\hat{v}^{3}i \\
\sum wv\hat{v}^{3}i/a^{2} & \sum wv\hat{v}^{3}i/a & \sum wv\hat{v}^{3}i^{2}/a^{2} & \sum wv\hat{v}^{3}i^{2}/a \\
\sum wv\hat{v}^{3}i/a & \sum wv\hat{v}^{3}i & \sum wv\hat{v}^{3}i^{2}/a & \sum wv\hat{v}^{3}i^{2}
\end{bmatrix}^{-1}
$$

$$(3.25)$$

in which the experimental variance is defined as $\sigma_{\mathrm{exp}}^{2} = SS/(n-4)$, where $SS = \sum w(v-\hat{v})^{2}$, n is the number of observations, and we subtract 4 rather than 2 because four parameters have been estimated. In practice one is more likely to be interested in the variances of the non-linear parameters of the original equations than in those of the linear parameters, but as each non-linear parameter is the ratio of two linear parameters (or the reciprocal of one), eqns 1.22 and 1.23 apply. For example, the variance of $\hat{K}_{ic}$ is given by

$$\sigma^{2}(\hat{K}_{ic}) = \sigma^{2}\left(\frac{\hat{b}_{1}}{\hat{b}_{3}}\right) \approx \frac{\hat{b}_{1}^{2}}{\hat{b}_{3}^{2}}\left[\frac{\sigma^{2}(\hat{b}_{1})}{\hat{b}_{1}^{2}} - \frac{2\,\mathrm{cov}(\hat{b}_{1},\hat{b}_{3})}{\hat{b}_{1}\hat{b}_{3}} + \frac{\sigma^{2}(\hat{b}_{3})}{\hat{b}_{3}^{2}}\right].$$

$$(3.26)$$

All of this may seem very complicated, especially as apparently the whole argument has to be repeated for each model to be fitted. In reality, however, it is all much more straightforward than it may seem, because computer algorithms for matrix inversion are readily available and standardized, and the same calculations apply to all appropriate models provided that they are done on the general linear model: this means that all of the substitutions into the linear form must be made as the first step, and all of the back-substitutions to recover the kinetic equation as the last step.

What, however, is an 'appropriate model'? For this general method to apply, the non-linear kinetic equation must be linear if written in reciprocal form. This means that the right-hand side of the rate equation must be a fraction with only one term in the numerator and a denominator that is a linear function of the parameters. Fortunately this definition includes virtually all of the equations of interest in steady-state enzyme kinetics (including pH-dependence equations), so much so that it is simpler to list the cases that cannot be handled than those that can. The only important exceptions are the following.

1. The equation for *pure non-competitive inhibition* (the same as eqn 3.20 with the constraint that the two inhibition constants are identical) cannot be handled, because there is no provision in the calculation for a constraint such as $\hat{b}_1 \hat{b}_4 \equiv \hat{b}_2 \hat{b}_3$.

2. Equations for rates measured under *reversible conditions* (i.e. in the presence of a complete set of substrates and products) are exceptions because such equations always contain two terms in the numerator, one each for the forward and reverse reactions.

Considering now the cases that can be handled, two are obviously just special cases of mixed inhibition: the analysis of *competitive inhibition* is just the same apart from omission of all terms in K_{iu} and b_4 and putting $n = 3$ instead of $n = 4$ as the number of parameters; that for *uncompetitive inhibition* is the same apart from omission of terms in K_{ic} and b_3, putting $n = 3$, and renumbering the indices so that term 4 in every case replaces the missing term 3. However, these are by no means the only important models that are of the right form. Table 3.1 lists the appropriate substitutions for the models of enzyme inhibition, two-substrate reactions and pH dependence, i.e. the principal cases with up to four parameters. Models with more than four parameters (such as those for three substrates, or two substrates and an inhibitor, etc.) can be dealt with in the same way, but are not listed in the table.

3.4 Comparing models

When statistical tests disagree with common sense the most likely explanation is that the statistical test has been done incorrectly.

Cornish-Bowden and Cárdenas 1987

The main objective in data analysis is often not parameter estimation but *model discrimination:* we may be interested in the value of an inhibition

Table 3.1 Relationship between common enzyme kinetic equations and the general linear model

Model	Equation	n	y	x_1	x_2	x_3	x_4	Weight	b_1	b_2	b_3	b_4
Michaelis–Menten	$v = \dfrac{Va}{K_m + a}$	2	a/v	1	a			$wv\hat{v}^3/a^2$	$\dfrac{K_m}{V}$	$\dfrac{1}{V}$		
Substrate inhibition	$v = \dfrac{Va}{K + a + a^2/K_{si}}$	3	a/v	1	a	a^2		$wv\hat{v}^3/a^2$	$\dfrac{K}{V}$	$\dfrac{1}{V}$	$\dfrac{1}{VK_{si}}$	
Competitive inhibition	$v = \dfrac{Va}{K_m(1 + i/K_{ic}) + a}$	3	a/v	1	a	i		$wv\hat{v}^3/a^2$	$\dfrac{K_m}{V}$	$\dfrac{1}{V}$	$\dfrac{K_m}{VK_{ic}}$	
Uncompetitive inhibition	$v = \dfrac{Va}{K_m + a(1 + i/K_{iu})}$	3	a/v	1	a	ai		$wv\hat{v}^3/a^2$	$\dfrac{K_m}{V}$	$\dfrac{1}{V}$	$\dfrac{1}{VK_{iu}}$	
Mixed inhibition	$v = \dfrac{Va}{K_m(1 + i/K_{ic}) + a(1 + i/K_{iu})}$	4	a/v	1	a	i	ai	$wv\hat{v}^3/a^2$	$\dfrac{K_m}{V}$	$\dfrac{1}{V}$	$\dfrac{K_m}{VK_{ic}}$	$\dfrac{1}{VK_{iu}}$
Ternary-complex mechanism	$v = \dfrac{Vab}{K_{iA}K_{mB} + K_{mB}a + K_{mA}b + ab}$	4	a/v	1	a	a/b	$1/b$	$wv\hat{v}^3/a^2$	$\dfrac{K_{mA}}{V}$	$\dfrac{1}{V}$	$\dfrac{K_{mB}}{V}$	$\dfrac{K_{iA}K_{mB}}{V}$
Substituted-enzyme mechanism	$v = \dfrac{Vab}{K_{mB}a + K_{mA}b + ab}$	3	a/v	1	a	a/b		$wv\hat{v}^3/a^2$	$\dfrac{K_{mA}}{V}$	$\dfrac{1}{V}$	$\dfrac{K_{mB}}{V}$	
Rising pH dependence	$k = \dfrac{k_{lim}}{1 + b/K}$	2	b/k	1	b	b^2		$wk\hat{k}^3/b^2$	$\dfrac{1}{k_{lim}}$	$\dfrac{1}{k_{lim}K}$		
Falling pH dependence	$k = \dfrac{k_{lim}}{1 + K/b}$	2	b/k	1	b			$wk\hat{k}^3/b^2$	$\dfrac{K}{k_{lim}}$	$\dfrac{1}{k_{lim}}$		
Bell-shaped pH dependence	$k = \dfrac{k_{lim}}{b/K_1 + 1 + K_2/b}$	3	b/k	1	b	b^2		$wk\hat{k}^3/b^2$	$\dfrac{K_2}{k_{lim}}$	$\dfrac{1}{k_{lim}}$	$\dfrac{1}{k_{lim}K_1}$	

constant for its own sake, but we are more likely to be interested in whether one model for inhibition accounts for the data better than another, because this tells us something about the mechanism of the reaction. Likewise, when we estimate how a parameter changes with the conditions, we are less likely to be really interested in the numbers than in what they tell us about the mechanism.

It might seem at first sight that it would be sufficient to know whether one model gives a smaller sum of squares than another to be able to say that it accounts for the data better. This is too simple, however, because the simpler models are often special cases of the more general ones: the Michaelis–Menten equation is a special case of competitive inhibition in which the inhibition constant tends to infinity; competitive inhibition is a special case of mixed inhibition in which the uncompetitive inhibition constant tends to infinity, and so on. A moment's reflection will show that a general model must always give a sum of squares no larger than that given by any of its special cases; consequently the sum of squares must always decrease (or in the extreme case remain unchanged) when the model is generalized by introducing extra parameters. Thus taking the model with the smallest sum of squares will always lead us to prefer the most complicated model we can find. If we are dealing with linear problems (and often even if we are not) we can always make the sum of squares zero by choosing a model that has as many parameters as there are observations. All of this is contrary to the principle of William of Occam, which underlies much of science and enjoins us to search for the simplest explanation consistent with the facts.

We can improve on the sum of squares as a criterion for comparing models by replacing it with the *mean square MS*, defined as the sum of squares divided by the number of degrees of freedom, the number of observations n minus the number of parameters p:

$$MS = SS/(n - p). \tag{3.27}$$

This is apparently the same quantity that we have previously called the *experimental variance* (for example in Section 2.4), but we avoid that term here because strictly it carries an implication that it has been calculated from the correct model, and for comparing models we must avoid any such implication.

The mean square is certainly better than the sum of squares as a criterion for comparing models, because the presence of $-p$ in the denominator prevents it from mindlessly decreasing whenever another parameter is added to the model. Nonetheless, it is insufficient, because to know whether a more complex model is really needed, we need some criterion to judge whether the decrease in MS resulting from the introduction of one or more extra parameters is big enough not to be ascribed to

Table 3.2 Inhibition equations

(a) Data.

Substrate concentration	Inhibitor concentration				
	0	1	2	3	4
1	2.36	1.99	1.75	1.60	1.37
2	3.90	3.35	2.96	2.66	2.35
3	5.30	4.40	3.98	3.58	3.33

(b) Results.

Equation	V	K_m	K_{ic}	K_{iu}	$\Sigma(v/\hat{v} - 1)^2$
$v = \dfrac{Va}{K_m + a}$	11.2 ± 4.6	5.0 ± 2.8			4.05859370
$v = \dfrac{Va}{K_m(1 + i/K_{ic}) + a}$	11.4 ± 0.7	3.80 ± 0.25	4.21 ± 0.21		0.07010008
$v = \dfrac{Va}{K_m + a(1 + i/K_{iu})}$	32 ± 16	14 ± 8		0.85 ± 0.47	0.54905197
$v = \dfrac{Va}{K_m(1 + i/K_{ic}) + a(1 + i/K_{iu})}$	12.9 ± 1.1	4.50 ± 0.55	5.20 ± 0.67	12.2 ± 7.1	0.05130968

chance; in statistical terminology we need to be able to assess the *significance* of the improvement.

To make the problem more concrete, let us examine the data shown in Table 3.2(a), where the rate of an enzyme-catalysed reaction is shown at all 15 combinations of three substrate concentrations a and five inhibitor concentrations i. Table 3.2(b) shows the parameter values obtained by minimizing the sum of squares of relative deviations for the three commonest inhibition models, competitive, uncompetitive, and mixed. It also shows the results of trying the Michaelis–Menten equation (i.e. ignoring the presence of inhibitor). Although in this example it is obvious from inspection of the raw data that the inhibitor has an effect, this may not always be as clear as it is here, and as a general rule one should never discard a priori the possibility that there is no effect to be observed.

When the results are examined as plots of a/v against a (Fig. 3.1), it is clear that the uncompetitive equation fits rather badly, insisting on a convergence to a point on the a/v axis that is not justified by the data. Discrimination between the other two inhibition models is much less clear, and one cannot easily judge whether the weak convergence evident in the fit to mixed inhibition is required by the data or not. No other ordinary plot could do much better, because the information about which models fit better is compressed into small differences in the deviations of the points from the lines. What is needed is to magnify these differences by plotting not the original observations but the *differences* between the observations and the calculated values, in other words plotting *residuals*.

There are many reasonable choices for both ordinate and abscissa axes in a residual plot, and as the different choices emphasize different aspects of the data, it is usually a good idea to examine as many different residual plots as one has time and energy to produce. However, to prevent this book from expanding to the point where it consists almost entirely of residual plots, it will be necessary to show some restraint. In publishing results of experiments even greater restraint will almost certainly be required by the journal, but no such restraint is needed in the privacy of the laboratory, and the ideal number of residual plots to make of any fit is almost unlimited. The acceptable minimum number is one, and any computer program that fails to display at least one residual plot for every fit (whether the user asks for it or not) is failing in its responsibilities.

If only one residual plot is to be made, this ought to be one to test whether the assumptions on which the fit has been made appear valid or not. If the correct model has been fitted with the correct weights, the weighted differences between observed and calculated rates, $w^{1/2}(v - \hat{v})$, should show no dependence on any of the variables in the model, whether dependent or independent. (The power $\frac{1}{2}$ appears in this expression because it is the *square* of the weighted difference that contributes to the sum of squares.) As one of the hypotheses commonly made about the

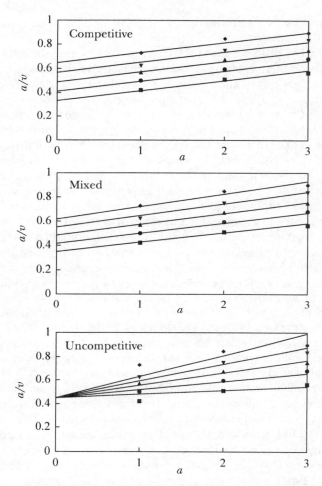

Fig. 3.1. *Primary plots of inhibition data.* The data consisted of the following set of (a, i, v) triplets: $(1, 0, 2.36)$, $(1, 1, 1.99)$, $(1, 2, 1.75)$, $(1, 3, 1.60)$, $(1, 4, 1.37)$, $(2, 0, 3.90)$, $(2, 1, 3.35)$, $(2, 2, 2.96)$, $(2, 3, 2.66)$, $(2, 4, 2.35)$, $(3, 0, 5.30)$, $(3, 1, 4.40)$, $(3, 2, 3.98)$, $(3, 3, 3.58)$, $(3, 4, 3.33)$. These are shown as plots of a/v against a with lines calculated from the best fits for competitive inhibition (*top*), mixed inhibition (*middle*), and uncompetitive inhibition (*bottom*).

weighting factor is that it varies with the true rate, the first choice for abscissa in the residual plots should be the calculated rate (which is, according to the hypothesis that the correct model has been correctly fitted, a better estimator than the observed rate of the true rate). Thus if only one residual plot is to be made, the appropriate choice is to plot $w^{1/2}(v - \hat{v})$ against $\hat{v}$. For the data of Table 3.2(a), which were fitted

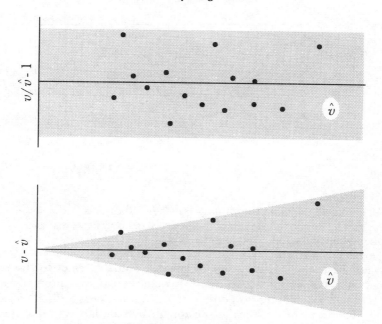

Fig. 3.2. *Residuals and weighting.* The results of Fig. 3.1 are replotted as residual plots of $v/\hat{v} - 1$ (*top*) and $v - \hat{v}$ (*bottom*) against $\hat{v}$, for calculated rates $\hat{v}$ obtained assuming mixed inhibition. This type of residual plot is useful for checking the appropriateness of the weighting scheme. When properly weighted residuals are plotted against the calculated value, the points should be scattered in a parallel band about the $\hat{v}$ axis, as in the plot at the top.

assuming relative errors, this is the same as $v/\hat{v} - 1$, and it may be seen in the upper part of Fig. 3.2 that the result is a uniform scatter of points in a parallel band about the horizontal axis, consistent with the hypothesis that the correct weights were used in the analysis. If $v - \hat{v}$ is plotted as ordinate instead of $v/\hat{v} - 1$ (lower part of Fig. 3.2), the result provides additional confirmation, as now the points are not scattered in a parallel band but a triangular band, indicating that the absolute magnitude of the differences increases with $\hat{v}$. Taken together, the two residual plots suggest that the relative weighting used was the right choice: if the data really followed simple errors the lower plot ought to show a horizontal band and the upper one to show convergence with increasing $\hat{v}$.

A complication (which does not apply to Fig. 3.2) can easily occur with large data sets if one or two observations have very large values of $w^{1/2}(v - \hat{v})$. Such observations can only be plotted on a linear scale if all the other observations are squashed so close to the axis that it becomes difficult to see whether there is any trend. One solution is to replot the remaining points with the outliers omitted, but this is dangerous as they

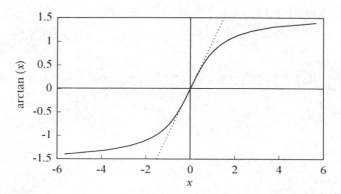

Fig. 3.3. *Stabilizing function for large numbers.* When x is in the range -0.6 to 0.6, arctan(x) (measured in radians) is within 10 per cent of x; but large, and even infinite, values are mapped into a finite range.

may not be outliers at all but may provide valuable clues to the correct model. A better solution, which is also convenient for automatic implementation in a computer program, is to use a non-linear ordinate scale that closely approximates a linear scale for all but the largest values. Such a scale is provided by the function arctan(x), which allows even infinite values to be plotted on a finite scale, but is virtually indistinguishable from x at low values (Fig. 3.3). As long as the $w^{1/2}(v - \hat{v})$ values are divided by a suitable scaling factor before conversion to arctangents, one can ensure that all but the largest deviations fall in the near-proportional range. For this purpose twice the mean of all the $w^{1/2}(v - \hat{v})$ values proves to give satisfactory results (the exact value is of no importance and others could be used, with almost no effect on the appearance of the final residual plot).

In this example the initial residual plot gives no clear indication that a wrong model has been fitted, but suppose that there were clear evidence of systematic error: how might we make additional residual plots to shed light on the reasons for the inadequacy of the model. To make this more concrete, suppose that the true model is mixed inhibition,

$$v_{\text{true}} = \frac{Va}{K_{\text{m}}(1 + i/K_{\text{ic}}) + a(1 + i/K_{\text{iu}})}, \tag{3.28}$$

but that the data have been fitted to the equation for competitive inhibition,

$$\hat{v} = \frac{\hat{V}a}{\hat{K}_{\text{m}}\big(1 + i/\hat{K}_{\text{ic}}\big) + a}. \tag{3.29}$$

The simple difference between the calculated and true rates is given by an inconveniently complicated function, but the difference between a/v values is straightforward:

$$\frac{a}{\hat{v}} - \frac{a}{v_{\text{true}}} = \left(\frac{\hat{K}_{\text{m}}}{\hat{V}} - \frac{K_{\text{m}}}{V} \right) + i \left(\frac{\hat{K}_m}{\hat{V}\hat{K}_{\text{ic}}} - \frac{K_{\text{m}}}{VK_{\text{ic}}} \right) + a \left(\frac{1}{\hat{V}} - \frac{1}{V} \right) - \frac{ai}{VK_{\text{iu}}}.$$

$$(3.30)$$

We cannot use this directly, because v_{true} is unknown, but if systematic error is predominant we may suppose that the observed rate v is a better estimator of v_{true} than the calculated rate $\hat{v}$, so we should be able to plot eqn 3.30 approximately by plotting $a/v - a/\hat{v}$ against a or i, and in either case the equation predicts a straight-line dependence. In practice, unless we have made a grossly bad choice of initial model, there will always be a substantial amount of random contribution to the residuals to complicate the analysis, but we may still hope to be able to deduce some useful information from a residual plot of this kind. For the data of Table 3.2 the results, shown in Figs. 3.4 and 3.5, are still rather equivocal about whether the data really require the equation for mixed inhibition. Although the largest residuals decrease substantially on passing from competitive to mixed, the improvement is not impressive. The proper conclusion, therefore, is that there is a weak suggestion that an uncompetitive term is needed in the rate equation, but the data fail to establish this. The next step, obviously, is to enquire *why* the data fail to establish the right model. The research is not finished when the analysis has been done! A good analysis should suggest what was wrong with the design of the experiment, so that one can go back and do it again with greater hopes of success. In Table 3.2 the obvious design failure is that the largest inhibitor concentration studied is only one-third of the estimated value of K_{iu} (it is less even than the estimated value of K_{ic}). It follows that satisfactory characterization of the inhibition requires a range of inhibitor concentrations that extends to considerably higher values.

It is tempting, when examining an equation such as eqn 3.30, to notice that the first three coefficients on the right-hand side are all differences between approximately equal numbers whereas the fourth is a new parameter, and to conclude that the first three terms in the expression ought to be negligible compared with the fourth. This temptation must be resisted, not only because any residual plot is a mechanism for magnifying small differences, but also because one cannot in general assume that adding a new parameter to a model will leave almost unchanged those already estimated. If the simpler model has been well chosen the last parameter to be added ($1/VK_{\text{iu}}$ in the example considered) will usually be the least

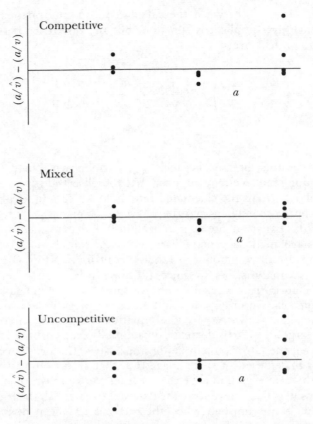

Fig. 3.4. *Residual plots of inhibition data.* The results of Fig. 3.1 are replotted in the form of residual plots of $(a/\hat{v}) - (a/v)$ against the substrate concentration a. This type of plot is useful for revealing whether there is any dependence on a unexplained by the fitted line.

important, and its value need not be large compared with the *changes* in the values of the others.

I have discussed residual plots in some detail before mentioning statistical tests, because I believe that residual plots represent the ideal, statistical tests the second best. True, statistical tests have a large body of theory behind them, and they allow the significance of a result to be expressed as a number, but the large body of theory is built on a large body of assumptions (see Chapter 4) that may have little or no validity in relation to the particular data we wish to analyse. Moreover, it is rare for a statistical test to lead to a conclusion that is not obvious to a careful observer. The main real advantage that tests have over graphs is that they can more easily be done fast and automatically in the computer; they may also look more impressive to the referees of your papers. Properly done,

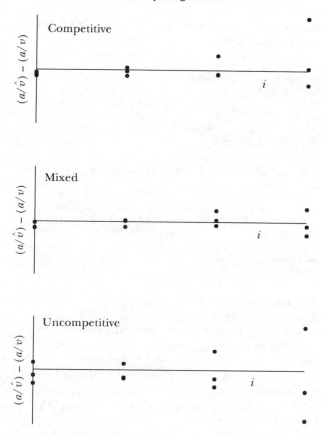

Fig. 3.5. *Residual plots of inhibition data.* The results of Figs. 3.1 and 3.2 are replotted in the form of residual plots of $(a/\hat{v}) - (a-v)$ against the inhibitor concentration i.

they are automated common sense; incorrectly done, they are an aid to self-deception.

The fact that a sum of squares is a sum means that it can be broken down into components: each of these has a number of degrees of freedom associated with it, and the components can be compared to decide whether they can reasonably be considered equivalent. Comparing the fits obtained with different models is an example of *analysis of variance* (often shortened to the acronym ANOVA), which I shall illustrate with the results given in Table 3.2(b), especially the comparison between competitive and mixed inhibition.

If competitive inhibition were the correct model, we should calculate the experimental variance as $0.070\,100\,08/12 = 0.005\,841\,73$, dividing by 12 because subtracting the three parameters from the 15 observations leaves

Table 3.3　Analysis of variance for comparing models

Hypothesis (a): Michaelis–Menten is the true model.

Source	SS	df	MS	F
Total (corrected for V, K_m)	4.05859370	13		
$K_{ic} \mid V, K_m$	3.98849361	1	3.98849361	682.766[a]
Residual	0.07010008	12	0.00584167	

[a] $P = 6 \times 10^{-12}$ (highly significant).

Hypothesis (b): Competitive inhibition is the true model.

Source	SS	df	MS	F
Total (corrected for V, K_m, K_{ic})	0.07010008	12		
$K_{iu} \mid V, K_m, K_{ic}$	0.01879040	1	0.01879040	4.028[a]
Residual	0.05130968	11	0.00584167	

[a] $P = 0.07$ (not significant).

12 degrees of freedom; whereas if mixed inhibition were the correct model we should calculate it as $0.05130968/11 = 0.00466452$. If competitive inhibition were really the correct model, then the extra parameter had no more than a chance effect on the fit, and so the particular component 0.01879040 of the total 0.07010008 represented by the decrease to 0.05130968 was no different from the rest. The *null hypothesis* (i.e. the hypothesis that will be tested), is that the mean square $0.01879040/1 = 0.01879040$ is not significantly different from the mean square $0.05130968/11 = 0.00466452$. The ratio of the two is $0.01879040/0.00466452 = 4.028$, and so the question to ask is how often a ratio this big would occur for a ratio of mean squares with 11 and 1 degrees of freedom if they were calculated from samples from the same distribution. The answer can be found in tables of the F statistic, and is about 7 times in 100 trials. Conventionally we take 5 per cent as the dividing line between 'significant' and 'not significant', and so we conclude that the improvement due to introduction of the parameter K_{iu} is not significant, so we cannot reject the hypothesis that competitive inhibition is the true model.

Analysis of variance calculations are commonly set out as a table, as illustrated for this example as Table 3.3(b), together with another example (Table 3.3(a)) given to illustrate the sort of results obtained when the improvement is clear: statistical tests are very effective for confirming the obvious, but less good for shedding light on difficult problems. In the top line of each part of the table, the qualification '(corrected for ...)' reflects the fact that although one may begin by examining *all* of the variation in the dependent variable, one normally omits the part that is regarded as too obvious to be given in detail. Thus, 'Total (corrected for V, K_m)' means all the variation in v that is left after the variation that can be explained by the Michaelis–Menten equation has been removed.

One should perhaps comment on the high precision with which the numbers are given in Table 3.3. It is conventional in analysis-of-variance calculations to give results to the full precision used in the calculation, though whether this really serves any purpose is arguable. Occasionally it may avoid problems of rounding error; more often it probably serves no purpose at all, and certainly in Table 3.3 nothing of importance would be lost if all sums of squares and mean squares were expressed to ± 0.0001.

The discussion in this section, and indeed throughout this chapter, uses numerical examples primarily to illustrate points of principle in data analysis. A much more data-oriented discussion of the analysis of enzyme inhibition may be found in Chapter 7.

3.5   Additional remarks about residual plots

Residual plots differ in two major respects from most plots used by scientists: they should be labelled with the *least possible amount of extraneous information*, and they do not need to be drawn with great care and accuracy. Figures 3.2–3.4 would be considered under-labelled for most purposes, with neither scales nor units indicated, but they are *over*-labelled from the point of view of their function: the names given to the panels and the labels on the axes are perhaps necessary for a published graph, but they distract from the visual message of the points. For interpreting a residual plot the exact values, units, variables plotted are irrelevent to the sort of question that one is trying to answer, and should be confined to a legend or the accompanying text, so that one sees the pattern of points about the axes, and *nothing else*. Even the axes could be regarded as redundant for some purposes (they are not needed for recognizing whether there is a systematic trend or not, for example), but it is usually helpful to have a visual indication of whether the values are positive or negative. If the plot combines data obtained in different ways (such as at different concentrations of an inhibitor), it may often be helpful to distinguish between them with different symbols for the points, as in Figs. 7.2–3.

The fact that one is mainly concerned with the visual impact of the plot means that highly accurate plotting serves little purpose. For example, the top panel of Fig. 3.6 is drawn from the same data as the top panel of Fig. 3.2, but the values were rounded to two significant figures in both coordinates before plotting: the two plots are essentially the same, giving exactly the same visual message, and it is quite difficult to detect any difference between them. This means that lack of graph paper provides no excuse for not making a residual plot—a hand-drawn plot on plain paper

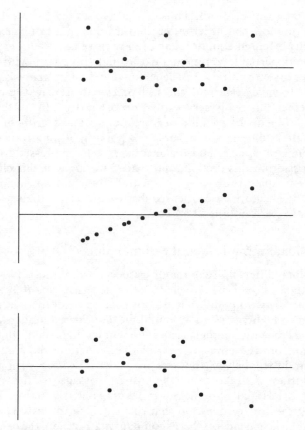

Fig. 3.6. *Different kinds of error in a residual plot.* Idealized examples are shown of plots of residual deviation against calculated value when the deviations are dominated by random error (*top*), systematic error (*middle*), or rounding error (*bottom*). In real examples the different kinds of error occur simultaneously in different proportions, so the results are less clear-cut than those shown here. Residual plots are primarily useful for their visual impact, and are best examined without any labelling to distract the eye.

is better than no plot at all—and the 80×25 resolution of a typical computer screen in text mode is perfectly adequate for a residual plot, however crude it may look for other purposes.

As mentioned already, the top panel of Fig. 3.6 shows the typical appearance of a residual plot with the calculated value as abscissa in the case where random error is the only important source of variation, and when the correct weights have been used. The other two complete the story by providing idealized examples of the two other ways in which differences between observed and calculated values can arise, *systematic error* and *rounding error*. They are idealized in the sense that in practice it is rare for either of these to be the *only* important source of variation: systematic and rounding effects will normally have some random character superimposed.

Systematic error is essentially the error that results from fitting the wrong model, and the whole process of data analysis can be regarded as one of recognizing and eliminating systematic effects. One might suppose that systematic effects as gross as those shown in Fig. 3.6 would rarely reach the published literature. Sadly, however, it is not difficult to find examples in current issues of journals of biochemistry. The main difference is that the published data are not usually presented in the form of residual plots—if they were the authors would have realized that their models did not fit the facts—but it is usually possible to assess what the residual plot would look like by looking along the line with the eye close to the paper.

In the past it would not have been thought necessary to give an example of the appearance of a residual plot where rounding error was the main source of error. Before high-performance spectrophotometers with built-in computing became available, experimenters were close enough to their data not to make the mistake of thinking their measurements much less precise than they were. Times change, however, and instruments now are available that are optically and photometrically capable of making measurements more than an order of magnitude more accurate than the numbers that appear on the print-out, and which offer no possibility either of making the estimates by hand or of increasing the precision in the print-out. When this happens the residual plots can be dominated by rounding error, as seen in the bottom panel of Fig. 3.6: a specific example is discussed in Cárdenas and Cornish-Bowden (1993). This is, of course, a problem that should never occur: it results from incompetent programming, not from any limit imposed by nature. Provided that there is direct access to the original observations rounding error can *always* be eliminated by more precise calculation.

3.6 Use of replicate observations

Most people are taught that it is a good idea to make replicate observations, but it is less often made clear *why* it is a good idea, or what information replicates provide that would be difficult to obtain in other ways. To this point we have treated the problem of discriminating between models as one of comparing at least two models. However, if the data contain replicates we can get an idea of whether the fit is adequate even if we have no other model available that might fit better. This is because replicates provide a model-independent measure of reproducibility. Generally speaking, with an acceptable model the replicates should agree with one another about as well as they agree with the model: the spread of values within a group of replicates should be neither very large nor very small on average in comparison with the dispersion about the fitted line.

If the replicates agree much better with each other than they do with the line, there are two possible reasons, and it is important to eliminate the first before leaping to the conclusion that the second is correct. The trivial interpretation is that the replicates are not true replicates, i.e. that they have more *reason* to agree with one another than with the rest. In making a replicate observation one must avoid making any 'convenient' simplification like thinking that because it requires stock solutions of the same concentration one can do the dilutions once only, or thinking that it is simpler to make a set of replicates one after another. The latter error, which results in unwanted time-dependent effects, is especially important to avoid in experiments with biological materials that deteriorate at a perceptible rate during the experiment. Ideally there should be no correlation at all between the values of the independent variables for the different observations and the order in which the observations are made; this applies in principle to all the observations, not just to groups of replicates.

Once one is satisfied that one's replicates are true replicates, one can assess whether they agree too well with one another. If they do, they imply that the model is inadequate and that a better fit would result from adding new terms to the fitted equation. This can be judged from a residual plot, or one can carry out an analysis of variance to compare the sum of squares due to *pure error* (a sum of squares calculated from the replicates alone) with the sum of squares due to *lack of fit* (what is left when the pure-error sum has been subtracted). Each group of m replicates makes a contribution of $\sum w(v - \bar{v})^2$, where $\bar{v}$ is the mean v within the group of replicates, to the pure error sum, and w must be the same weight as has been used for v in fitting the equation. The contribution has $m - 1$ degrees of freedom (because one degree of freedom has been 'used' for calculating the mean); the total sum of squares for pure error is the sum of all the individual sums for the different groups of replicates, and its number of degrees of freedom is the sum of all the individual numbers. The sum of squares due to lack of fit is then calculated by subtracting the sum for pure error from the total sum of squared deviations from the model, and its number of degrees of freedom is calculated in the same way. One can then make an F test of equality between the two mean squares.

Table 3.4 shows an example of how the calculation is done. As F values less than 1 are never significant, no statistical tables are needed to conclude that the value of $F = 0.609$ indicates that lack of fit is not significant in this example, i.e. the replicate observations provide no grounds for rejecting the model (which is not the same as saying that the model is therefore correct!). If an F value is only a little less than 1, as in this case, there is nothing more to be said. However, if it is much less than 1, so that for example $1/F$ would have been significant if the test had

Table 3.4 Testing for lack of fit

(a) Observed, mean, and calculated rates. The data were fitted to the equation $v = Va/[K_m(1 + i/K_{ic}) + a]$ assuming relative errors ($w = 1/\hat{v}^2$). The best-fit parameter values were $\hat{V} = 9.813\,55$, $\hat{K}_m = 3.104\,81$, $\hat{K}_{ic} = 3.605\,62$.

a	i	$w = 1/\hat{v}^2$	v	$\bar{v}$	$10^4 w(v - \bar{v})^2$	$\hat{v}$	$10^4 w(v - \hat{v})^2$
1	0	0.174959	2.45	2.38500	7.39199	2.39074	6.14410
1	0	0.174959	2.32	2.38500	7.39199	2.39074	8.75518
2	0	0.067647	3.89	3.89000	0.00000	3.84483	1.38021
3	0	0.042998	4.78	4.80000	0.17199	4.82253	0.77775
3	0	0.042998	4.88	4.80000	2.75188	4.82253	1.42014
3	0	0.042998	4.74	4.80000	1.54793	4.82253	2.92869
4	0	0.032759	5.47	5.47000	0.00000	5.52502	0.99168
5	0	0.027831	5.97	5.95500	0.06138	6.05415	1.93197
5	0	0.027831	5.94	5.95500	0.06138	6.05415	3.55504
1	1	0.256063	1.96	1.96000	0.00000	1.97618	0.67035
2	1	0.092394	3.28	3.28500	0.02309	3.28987	0.09000
2	1	0.092394	3.29	3.28500	0.02309	3.28987	0.00001
3	1	0.055984	4.30	4.30000	0.00000	4.22639	3.03343
4	1	0.041181	4.98	4.99000	0.04118	4.92777	1.12341
4	1	0.041181	5.00	4.99000	0.04118	4.92777	2.14849
1	2	0.352564	1.66	1.70500	7.13942	1.68415	2.05623
1	2	0.352564	1.75	1.70500	7.13942	1.68415	15.28797
2	2	0.120990	2.78	2.78000	0.00000	2.87492	10.90093
3	2	0.070680	3.90	3.79000	8.55227	3.76142	13.57367
3	2	0.070680	3.68	3.79000	8.55227	3.76142	4.68552
4	2	0.050566	4.52	4.52000	0.00000	4.44705	2.69095
5	2	0.040110	5.08	4.99000	3.24889	4.99315	3.02545
5	2	0.040110	4.90	4.99000	3.24889	4.99315	3.48029
1	3	0.464468	1.44	1.44000	0.00000	1.46731	3.46417
2	3	0.153437	2.59	2.57750	0.23974	2.55291	2.11077
2	3	0.153437	2.61	2.57750	1.62067	2.55291	5.00091
2	3	0.153437	2.48	2.57750	14.58606	2.55291	8.15648
2	3	0.153437	2.63	2.57750	4.22909	2.55291	9.11853
3	3	0.295106	3.44	3.44000	0.00000	3.38861	2.29992
4	3	0.060912	4.01	4.01000	0.00000	4.05179	1.06377
5	3	0.047447	4.70	4.70000	0.00000	4.59087	5.65064
1	4	0.591771	1.32	1.28000	9.46832	1.29994	2.38130
1	4	0.591771	1.27	1.28000	0.59177	1.29994	5.30465
1	4	0.591771	1.25	1.28000	5.32593	1.29994	14.75878
2	4	0.189732	2.39	2.39000	0.00000	2.29578	16.84324
3	4	0.105206	3.09	3.03000	3.78742	3.08304	0.05096
3	4	0.105206	2.97	3.03000	3.78742	3.08304	13.44330
4	4	0.072222	3.68	3.68000	0.00000	3.72105	1.21701
5	4	0.055400	4.16	4.20500	1.12185	4.24858	4.34694
5	4	0.055400	4.25	4.20500	1.12185	4.24858	0.00111
Sum					103.268 36		185.863 94

(b) Analysis of variance.

Source	SS	df	MS	F
Total (corrected for V, K_m, K_{ic})	0.018586394	37		
Lack of fit	0.008259558	21	0.00039331	0.609[a]
Pure error	0.010326836	16	0.00064543	

[a] Not significant.

been done the other way around, it would suggest that a mistake had been made in the calculation, because very small F values are not expected to occur by chance any more often than very large values.

The calculation in Table 3.4 is useful for illustrating exactly what is gained by making replicate observations and how it is paid for. Of 40 original degrees of freedom contained in 40 observations there is no choice but to pay three to obtain the three parameters of the fitted equation. Of the 37 remaining degrees of freedom, one could use all 37 to measure lack of fit, but in this case there would be no replicate observations and hence no way of gauging the error level independently of the fitted model. By choosing to make 16 observations that replicate others, one gains independent information about the pure error, but nothing in this world comes without a price, and in this case 16 degrees of freedom for lack of fit are lost. In practice, therefore, one must compromise: too few replicates and one has no useful information about pure error; too many and one sacrifices detail about lack of fit. Put somewhat differently, it might seem at first sight that the whole of the sum of squares 0.018 586 394 is telling us something about how well the equation for competitive inhibition agrees with the data, but the analysis of variance makes it clear that this is wrong: only 0.008 259 558 is telling us anything about this; the rest is telling us something different, namely how well the replicates agree with one another.

4
Maximum likelihood and efficiency

Since this [the probability distribution] cannot be known a priori, we will, approaching the subject from another point of view, inquire upon what function, tacitly, as it were, assumed as a base, the common principle, the excellence of which is generally acknowledged, depends. It has been customary certainly to regard as an axiom the hypothesis that if any quantity has been determined by several direct observations, made under the same circumstances and with equal care, the arithmetical mean of the observations affords the most probable value, if not rigorously, yet very nearly at least, so that it is always most safe to adhere to it.

Karl Friedrich Gauss 1809

4.1 The theoretical basis of least squares

A professional statistician would not, perhaps, be very happy with the title of this chapter, because maximum likelihood and efficiency are by no means the same thing, and a treatise on statistics would normally deal with them separately. They belong together in this book because they constitute two sorts of arguments that are often advanced in favour of the method of least squares, and both of these are based on misconceptions that I want to try to clear up in this chapter. Neither is very important from the practical point of view (unless they are so persuasive that they discourage the experimenter from looking any further than least squares), and many readers will want to skip this chapter, apart perhaps from this introductory section, which sets out the main points in a non-technical way.

The first common idea is that one can prove in a general way that least-squares estimates are the best possible, where 'best', if it is defined at all, is taken to mean 'minimum variance'. In fact, the method of least squares, when properly weighted, does lead to minimum-variance estimators if the selection is limited to a particular class of estimators that are easy to calculate, and with certain assumptions it leads to minimum-variance estimators even when all conceivable methods are allowed, but it is not a minimum-variance method *in general*. This point is taken up in Section 4.2.

Another argument often advanced in favour of least-squares estimates is that they are equivalent to *maximum-likelihood* estimates. This is said by people who have only the vaguest idea of what maximum-likelihood estimates are—thinking perhaps that they are the estimates most likely to be the correct ones. As with efficiency, maximum-likelihood is a technical term with a definition I shall give below, one too technical to be summarized in a word or two, though I emphasize immediately that it does not have the naive meaning just referred to. Also, as with minimum-variance, least-squares estimates are not in general identical with maximum-likelihood estimates; they only become identical under certain assumptions, as will be discussed in Section 4.3.

The remainder of this chapter will be devoted to justifying the assertions that have been made in this section.

4.2   Minimum variance

It might seem from a superficial reading of Section 1.4 that we have already proved that a properly weighted least-squares estimator is a minimum-variance estimator, but this is not the case. First of all, defining $\hat{b} = \bar{b}$ in eqn 1.4 was done to make the expression for the sample variance a minimum, but when we talk about a minimum-variance estimator we are not concerned with the sample variance, but with the true variance.

Later, in Section 1.5 it might have seemed that this objection was taken care of in the derivation of eqn 1.27, which defined the proper weights necessary for minimizing the true variance of $\hat{b}$. The problem here is that the expression for the variance of $\hat{b}$ used eqn 1.8, i.e. it assumed that we were dealing with an estimator that could be expressed as the sum of a set of values of known variance; more generally, it assumed that the estimator was a linear function of the observed values, i.e. that it was a linear estimator. Among linear estimators, least-squares estimators are indeed minimum-variance estimators, but they are not the only possible kinds of estimator, and the derivation of eqn 1.27 says nothing about the possibility that other (non-linear) estimators might lead to lower variance than $\hat{b}$.

If we are willing to make some assumptions about the distribution of the observations, we can go further than this. Thus, if they follow a normal distribution, to be defined in the next section, it can be shown (though it will not be shown here) that the absolute minimum variance that any estimator can have is the same as that of the least-squares estimator: thus in this case the least-squares estimator is not just the minimum-variance linear estimator, it is also the minimum-variance estimator out of all possible estimators.

All of this would be pedantic if all the estimators we were likely to want to use were linear estimators, or if all observations in practice followed the normal distribution to an adequate approximation. However, neither of these is the case: among simple estimators, the median is not a linear function of the observed values, and when the true distribution is 'long-tailed' (by no means a very unlikely circumstance) rather than normal, the median can have a smaller variance than the mean. To take this point further, however, we need a definition of a normal distribution, which will now be given.

4.3 The normal distribution

Un physicien éminent me disait un jour à propos de la loi des erreurs: 'Tout le monde y croit fermement parce que les mathématiciens s'imaginent que c'est un fait d'observation, et les observateurs que c'est un théorème de mathématiques.'

Henri Poincaré 1892

Why does everyone believe in the normal distribution? A century ago, a physicist colleague of Poincaré thought he had the answer: mathematicians thought that it has been found true by experiment, experimentalists that it could be proved mathematically. The world has not much changed since then.

As will be shown shortly, an assumption that observations follow the normal distribution leads to the conclusion that the arithmetic mean is the most efficient ('best') estimate. It is quite easy to follow the argument in this direction, but Gauss, in fact, did something much more difficult when he derived the equation for the normal distribution (often, for that reason, called the 'Gaussian distribution'). He put the question exactly the opposite way round: he *assumed* that the arithmetic mean was best, and then deduced the underlying statistical properties of the data.

The quotation at the beginning of this chapter would not win any prizes in a literary competition, but its meaning is clear enough if one disentangles the stylistic contortions that the translator took such pains to preserve. Noting that the real probability function cannot be known a priori, Gauss decides to turn the question around: first he will choose a conclusion whose excellence is universally recognized; then he will establish what assumptions it follows from. As it is axiomatic that the arithmetic mean provides the most probable value, the problem is to find a distribution from which one can prove that this mean is best. Thus far from proving that the arithmetic mean (the archetype of a least-squares estimate) was best, Gauss defined a world in which it would be the best.

Suppose now that we lived in Gauss's world: what would it mean? It would mean that the probability that we would observe a value in the range x_1 to x_2 of a quantity with true value μ and variance σ^2 would be given by the area under the curve between $x = x_1$ and $x = x_2$ for the *normal distribution curve*, which is illustrated in Fig. 4.1 and is defined by the following equation:

$$f(x) = \frac{1}{\sigma\sqrt{2\pi}} \exp\left[\frac{-(x-\mu)^2}{2\sigma^2}\right]. \tag{4.1}$$

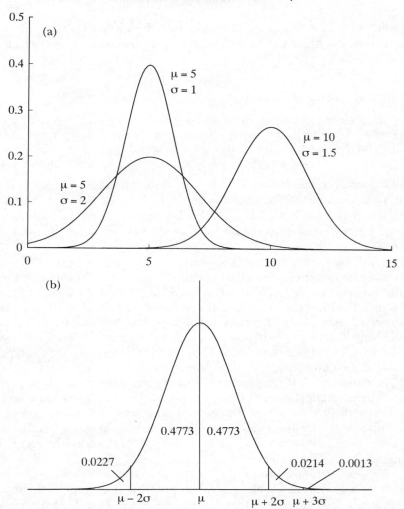

Fig. 4.1. *The Gaussian ('normal') distribution curve.* (a) The curve is a symmetrical bell, with a maximum at the mean μ and a width that increases with the standard deviation σ. (b) Errors less than 2σ occur with 95.46 per cent probability; errors less than 3σ occur with 99.74 per cent probability.

The area under the curve between $x = \mu - 2\sigma$ and $x = \mu + 2\sigma$ is twice 0.4773, or 0.9546, and that between $x = \mu - 3\sigma$ and $x = \mu + 3\sigma$ is twice (0.4773 + 0.0214) or 0.9974: this is the basis for the common idea that only 5 per cent of observations deviate from the mean by more than twice the standard deviation and that virtually all deviate by less than three times the standard deviation.

If we ask what will be the probability of observing a particular value x_i,

Table 4.1 Representative values for the
normal distribution

x	$f(x)$
μ	0.3989
$\mu + \sigma$	0.2420
$\mu + 2\sigma$	0.0540
$\mu + 3\sigma$	0.0044

the answer will always be zero, because this corresponds to a slice of zero width with $x_1 = x_2 = x_i$, but this need not present a problem as long as we talk about thin (but not infinitely thin) slices and we simply compare one x value with another. For example, if we compare $x_i = \mu + \sigma$ with $x_i = \mu + 2\sigma$, we see from Table 4.1 that the ratio of $f(x)$ values is 0.2420/0.0540, or 4.5, so that, for uncertainties of the same width in both cases, we should observe values one standard deviation away from the true value about 4.5 times more often than values two standard deviations from the true value. We could say that for a known mean and standard deviation an observation of $\mu + \sigma$ is 4.5 times more *likely* than one of $\mu + 2\sigma$.

If we have several different observations, we can do the same calculation for each one, and if they are independent observations we can multiply the individual values together to obtain a value for the *likelihood L* of the complete set of n observations:

$$L = \left(\frac{1}{\sigma\sqrt{2\pi}} \right)^n \prod \exp\left[\frac{-(x_i - \mu)^2}{2\sigma^2} \right]. \tag{4.2}$$

To take a concrete example, but using simple numbers so that we can refer directly to Table 4.1, suppose that $\mu = 0$, $\sigma = 1$ and we want to know the likelihood of obtaining four observed values, 0, 1, 2, and 3. It is evidently the product of the four numbers in the right-hand column, i.e. $L = 2.29 \times 10^{-5}$.

This example is, however, very abstract and unrealistic, because it supposes that we know what the mean and standard deviation are, and that we are interested in calculating the likelihood of a particular set of observations. The reality is likely to be almost the opposite: we do not know what the mean and standard deviation are, but we *do* know what values we have observed. The method of *maximum likelihood* then adopts as a principle that we should define our best estimates of μ and σ as those that maximize the likelihood of the particular observations that we have. How do we translate this principle into a usable method? First, we note that the form of eqn 4.2 is such that it will be much easier to work with the logarithm of L, but as $\ln L$ is a monotonic increasing function of L it is clear that if we maximize $\ln L$ then we maximize L. So, we have

$$\ln L = -\frac{n}{2}\ln \pi - n \ln s - \frac{1}{2s^2} \sum (x_i - m)^2, \qquad (4.3)$$

where m and s are arbitrary estimates of μ and σ respectively. It is obvious from inspection that for any s we can maximize $\ln L$ by minimizing the ordinary sum of squares of $(x_i - m)$ values. Thus the maximum-likelihood estimate of the mean is the least-squares estimate, which, as we have seen, is just the sample mean $\bar{x}$. What about s, however? Partially differentiating $\ln L$ with respect to s, we have

$$\frac{\partial \ln L}{\partial s} = -\frac{n}{s} + \frac{1}{s^3} \sum (x_i - \bar{x})^2, \qquad (4.4)$$

and it is again obvious from inspection that this is zero if s^2 is defined as the sample standard deviation $(1/n)\sum (x_i - \bar{x})^2$.

We appear to have shown, therefore, that maximum-likelihood estimation leads to the same results as least-squares estimation. Note, however, that we showed this only by starting with the assumption that the errors were normally distributed and, as we have seen, Gauss arrived at the equation for this distribution by assuming that this was the conclusion he wanted to arrive at. If he had started out with the idea that 'the median of the observation affords the most probable value, if not rigorously, yet very nearly at least, so that it is always most safe to adhere to it,'* then he would have defined the 'normal' distribution in a different way, and maximum likelihood would not have turned out to be the same as least squares.

4.4   How 'normal' is the normal distribution?

Normality is a myth; there never was, and never will be, a normal distribution.
                                                                    Geary 1947

The typical distribution of errors and fluctuations has a shape whose tails are longer than that of a Gaussian distribution.
                                                    Tukey and McLaughlin 1963

It is, perhaps, commoner than one might wish in research to decide what the conclusions are going to be before doing the research, often without being as explicit about it as Gauss was. Most would agree that this is not a good way to proceed, however, and to know how firmly based is the

---

*In reality the median is 'safer' than the mean, because it is much less affected by failures of the other assumptions that Gauss made.

statistical theory built on the idea of the normal distribution, we need to know how well it defines the distribution of real observations. This is, unfortunately, much more difficult to establish than one might think, and it has rarely even been attempted. The problem is that determining a distribution curve experimentally, even at a rather low level of accuracy, requires a huge number of observations, because one needs to know whether rare events are really as rare as the distribution function predicts. For example, according to the normal distribution, about one observation in 1000 should deviate from the mean by more than 3.3 times the standard deviation, and to make even a rough estimate of how frequently such 'once in a thousand' events really occur one needs to make several thousand observations of the same value—difficult enough for any sort of observation, and almost impossible for a typical enzyme assay in a spectrophotometer.

To make this more concrete, suppose we have two sets of values, one taken from a true normal distribution with mean 10 and standard deviation 1, the other from a distribution in which each observation has a 0.95 probability of coming from the same normal distribution, but a 0.05 probability of coming from a different normal distribution with mean 10 and standard deviation 3. The two curves are illustrated in Fig. 4.2, and are very similar in appearance, so similar in fact that few would argue if told that the curve in Fig. 4.2(b) was a normal curve; only an observer with some experience would easily recognize that the area under the tails was too large.

Nonetheless, similar as the two curves are, the difference between them is not trivial. It is sufficient, for example, to eliminate most of the advantage that the normal distribution gives to the mean over the median as an estimator. This is because the almost unnoticed tails are responsible for a large part of the total variance in the second case, which is 95 per cent of 1 added to 5 per cent of 3^2, or 1.4: about one-third of the total variance comes from the 5 per cent of outliers.

When it is claimed, therefore, that the mean should be preferred as an estimator over the median, on the grounds that it leads to a lower variance, the underlying assumption is that it is quite certain that the distribution curve looks like Fig 4.2(a) and not like Fig. 4.2(b)! As we can see from the figure, this is not so easy to be sure of even when one sees the smooth calculated curves. In reality one would be working with a histogram derived from a few tens of observations at most, which would give only a crude indication of the distribution.*

---

*To be fair, statistics textbooks describe methods such as *probit* and *rankit* plots for investigating distributions that are a bit more sensitive than drawing histograms. However, these cannot escape from the fundamental difficulty that there are never anywhere near enough data to determine the true distribution curve, not even the middle 99 per cent of it, i.e. the least important part of it.

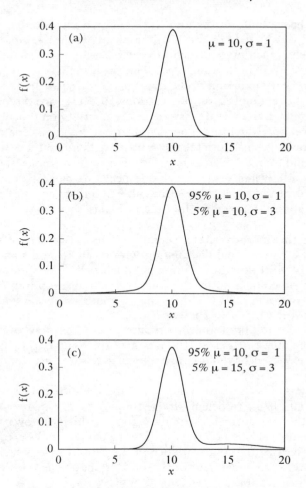

Fig. 4.2. *Contaminated normal curves*. (a) A true normal curve with mean 10 and standard deviation 1; (b) the same normal curve mixed with 5 per cent of a second normal curve with the same mean and standard deviation 3; (c) the same normal curve mixed with 5 per cent of a normal curve with mean 15 and standard deviation 3.

Outlying observations can arise from numerous causes, often unidentified, and it is probably more realistic to suppose that not only the standard deviation is perturbed, but the mean also. Figure 4.2(c) illustrates such a case, where the outliers, still at 5 per cent frequency, now have not only a standard deviation elevated by a factor of three, but also a mean that is 50 per cent too high. Although the calculated curve is more obviously non-normal than the one in Fig. 4.2(b), the difficulty of recognizing it *in practice* from a crude histogram remains, and its effects on the estimation are much more devastating. Although the outliers raise the true mean by only 2.5 per cent, from 10 to $0.95 + 0.05 \times 15$, or 10.25, they raise the variance by almost 160 per cent, from 1 to $0.95 \times 1^2 + 0.95 \times (10 - 10.25)^2 + 0.05 \times 3^2 + 0.05 \times (15 - 10.25)^2$, or 2.587.

4.5 Efficiency

Returning to the definition of the variance of a least-squares estimator $\hat{b}$ given in Section 1.4, we recall from eqn 1.12 that it is inversely proportional to the number of observations:

$$\sigma^2(\hat{b}) = \frac{1}{n} \sigma^2(b). \qquad (4.5)$$

Although the exact form of the relationship will vary according to the sort of estimator we are concerned with, the proportionality to $1/n$ is general. As the amount of effort required to do an experiment increases monotonically (sometimes even proportionally) with the number of observations made, this relationship provides a convenient 'economic' basis for comparing estimators and for expressing the idea of minimum variance in terms of costs, because it allows us to pose the question in the following way: if I know I must make 20 observations to obtain an adequately accurate value of the parameter I am interested in if I use statistical method A for analysing the data, how many observations must I make to obtain the same final accuracy if I use method B, which for equal n gives a 10 per cent higher variance in the parameter? The answer, evidently, is 10 per cent more than 20, or 22. We can then say not only that method B is less *efficient* than method A, in the sense that it requires more work to achieve the same final result, but also we can quantify it as having 20/22 or 91 per cent of the *efficiency* of method A.

We can get an idea of how this works out in practice by studying the relative efficiencies of the sample mean and median as estimators of the true mean of a population. The detailed theory of the distribution of the sample median is too advanced for this book, but we can take the results of Hojo (1931), discussed in Kendall and Stuart (1969), on trust.* These show that for normally distributed observations the limiting efficiency of the median is 63.7 per cent of that of the mean, implying for example that one needs to make 1.57 times as many observations to obtain the same final accuracy, and many have concluded from this that the median is too 'expensive' in terms of experimental effort to be regarded as an acceptable alternative to the mean.

*We do *not* have to take the qualitative behaviour on trust, as it is quite easy to show by computer simulation that Hojo's numbers are correct to within sampling variation.

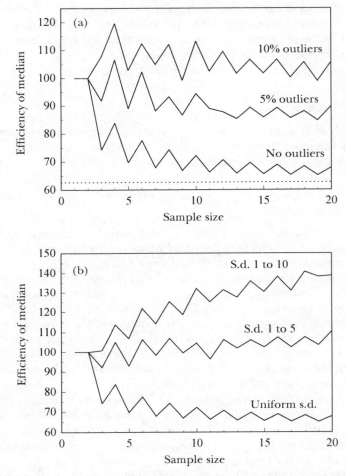

Fig. 4.3. *Relative efficiency of the median.* The lines show how the efficiency of the sample median compared with the sample mean as an estimator of the population mean varies with sample size, for various assumptions about error distribution. In all cases the zig-zag pattern derives from the different definitions of the sample median for odd and even numbers of observations (as a unique value or as the mean of two values respectively). The more regular patterns for the two Gaussian cases derive from the fact that these lines were calculated exactly from the known distribution of the sample median, whereas the others are more irregular because they were obtained by Monte Carlo sampling. Panel (a) shows the effect of a small proportion of outliers in an otherwise normal sample. Panel (b) shows the effects of making incorrect assumptions about weighting when calculating the sample mean.

There are three points to be made, however. First, the value of 63.7 per cent is a *limit* that assumes an infinite sample: biochemists do not habitually work with infinite numbers of observations; on the contrary, kinetic experiments are typically done with samples that are often very small by the standards of statistical theory. As Fig. 4.3 illustrates, the efficiency of the median for finite samples is always greater than 63.7 per cent, and although it decreases quite fast as the sample size increases, 70 per cent is a more realistic 'typical' value. Even 63.7 per cent efficiency may seem surprisingly high to people who take the simple-minded view that calculating a median involves 'throwing away data' whereas calculating the mean 'uses all the data'. In reality, of course, *all* of the observations contribute to the decision as to which observation (or the mean of which pair of observations) is the median, and none are 'thrown away'. One must always be suspicious of arguments about preferring methods that 'use all the data', because such arguments are usually a substitute for thinking clearly (for a different kind of example, see Cornish-Bowden, 1983); methods have to be judged by their results, not by pseudo-philosophy.

The second point is much more important: the calculation of the efficiency assumes that the underlying distribution is normal, and as the mean is much more sensitive to outliers than the median, even quite small departures from normality create much greater havoc when the mean is used than when the median is used. This is also illustrated in Fig. 4.3(a): with only 5 per cent outliers having standard deviation three times that of the 'good' observations, corresponding to the distribution curve of Fig. 4.2(b), the median is almost as efficient as the mean, and when the proportion of outliers is raised to 10 per cent it becomes clearly better than the mean.

The last point is perhaps the most important of all. Our calculation of the ideal efficiency of the median has assumed not only that the observations were normally distributed, but also that each observation came from the *same* normal distribution, i.e. that it had the same variance. If this were not the case, the calculation of the mean could always, in principle, be optimized by means of weights, *provided that there was information about what weights to use*. This is a crucial proviso, because although it may be reasonable to assume that observations of the same value made under the same conditions have uniform variance, it is by no means equally reasonable to make the same assumption for a sequence of observations of different values, such as a set of enzyme rates. If these vary over a factor of five and have uniform coefficient of variation, their standard deviations will span a fivefold range, but one may still use the median and mean to illustrate the sort of effects this may have by supposing that the observations have standard deviations that vary arbitrarily over a fivefold range. As Fig. 4.3(b) shows, the unweighted mean is no better than the median in this case, and when the unknown standard deviations span a tenfold range, the unweighted mean becomes definitely less efficient than the median.

The median is the simplest of a class of estimators known as *robust estimators*, which means they are more resistant than classical estimators to departures from the underlying assumptions. The more general problem, which has greater importance for enzyme kinetics than comparing medians and means, is how to develop methods of *robust regression*, i.e. methods of data fitting that are more resistant than least squares to failure of the assumptions. Chapters 5 and 6 are devoted to the discussion of such methods, but first we shall examine one of the arguments sometimes advanced by optimists for believing that we live in a Gaussian universe, and afterwards the chapter will conclude with a brief review of the assumptions that underlie the least-squares method.

4.6   The central limit theorem

Suppose that any particular error ε is composed of numerous components that contribute additively to the total, i.e.

$$\varepsilon = \varepsilon_1 + \varepsilon_2 + \varepsilon_3 + \cdots + \varepsilon_n. \tag{4.6}$$

Suppose, moreover, that each of these is distributed in an unknown way that deviates arbitrarily from a normal distribution, and that the only restriction that exists on their distributions is that they have *finite variances*. What can we then say about the distribution of ε? It turns out that there is a theorem in statistics called the *central limit theorem* that states that as n approaches infinity the distribution of ε approaches a normal distribution, regardless of the underlying distributions of the components.

If I thought that this mattered, I would then feel obliged to give some indication of the theoretical basis of the central limit theorem, and this would present a problem, as this theory is too advanced for a book of this kind. Fortunately, however, I think that the central limit theorem is an elegant irrelevance that has little or nothing to do with the real world of data fitting, and it will be sufficient, therefore, to indicate why the ordinary experimentalist can forget about it.

Although the assumptions that lead to the central limit theorem are superficially plausible, they contain an implausible element that undermines the whole edifice. While it may be true that real errors are composed of many components, it is manifestly *not true* that these are all of comparable importance, or even that there are several components of comparable importance. In typical 'analysis of variance' calculations, *one* source of error usually accounts for a large fraction of the total variance, *one* other accounts for a large fraction of what is left unaccounted for by the first, *one* other accounts for a large fraction of what is left unaccounted for by the first two, and so on. In these circumstances, the distribution curve of the total error is mainly accounted for by the main source of error, and if this is not normal then the total error will not be either. True, the presence of a large number of minor components may make it a *bit* more normal than the main component, but who cares?

In reality, if you want to convince yourself that you are dealing with normally distributed errors you must make a million or so repeated observations of the same thing under identical conditions and *measure* the distribution. The central limit theorem is a false trial that leads to nothing useful.

4.7 Review of assumptions implicit in least squares

The least-squares method, for which the estimator in the simplest kind of problem is the arithmetical mean, has the following properties.

1. Out of all linear estimators, the correctly weighted least-squares estimator has *minimum variance*, which is the same as saying that it has *maximum efficiency*.

2. If the observations are drawn from a normal distribution, the correctly weighted least-squares estimator has minimum variance out of all conceivable estimators, not just linear ones.

3. If the observations are drawn from a normal distribution, the least-squares estimator is a *maximum-likelihood* estimator.

Before we get too excited about these desirable properties, however, we should remember the following points.

1. Not all estimators of interest are linear estimators; specifically, the median is not a linear estimator, and robust estimation in general is not linear estimation.

2. We never know in practice whether the observations come from a normal distribution, and we rarely have any theoretical reason to expect that they do.

3. Even when we do have good reason to believe that the data are normally distributed, we rarely have such good information about the proper weights that we can be sure that we shall get the optimal behaviour that least squares is supposed to give us.

To get all the optimal behaviour, the following assumptions must be true.

1. The right model, i.e. the right kinetic equation, is being fitted.
2. The correct weights are known, which means that the function that defines the variance is known.
3. The errors in the observations are independent of one another.
4. They are normally distributed.
5. They are unbiassed, i.e. the errors have zero mean.

5
Generalized medians: looking beyond least squares

5.1 Doing without information on distributions and weights

As we have seen in Chapter 4, there are good reasons for regarding the method of least squares as a good method if the errors in the observations are known to be normally distributed and if the way in which the variances of these errors vary is also known, so that the data can be properly weighted. More generally, for any known distribution of errors and known weights there exists in principle a method ideal for that distribution. For the experimentalist at the bench, however, this is quite unreal. One normally knows nothing about the distribution curve, and although a reasonable guess at the proper weighting scheme can be made if one has a large number of points and takes the time to make residual plots, one also knows nothing *a priori* about the proper weights.

Very few studies have ever been carried out to determine the distribution or variances in real enzyme kinetic experiments. Of the few that exist only one (Burk 1934) suggests that it is correct to assign equal weight to all rates, and as this work is now very old it is of questionable relevance to modern experimental practice. More recent studies (Siano *et al.* 1975; Storer *et al.* 1975; Askelöf *et al.* 1976; Nimmo and Mabood 1979) suggest that relative errors may often be closer to the truth, and this truth may anyway be considerably more complicated than any dependence on v alone can represent (Mannervik *et al.* 1986).

Because of the large amount of data needed to establish a distribution, there is even less experimental information about whether it is realistic in practice to assume a normal distribution in enzyme kinetics, though several of the same papers describe efforts to learn something about it. None of them provides overwhelming evidence against assuming a normal distribution, but of course all of these studies are open to a theoretical objection: a realistic study of error distribution ought to be carried out in exactly the same way as an ordinary experiment, with exactly the same amount of care to avoid 'mistakes' and other causes of outliers; but this is virtually impossible to achieve in an experiment designed to study distribution, because real experiments do not consist of measuring the same rate many times. The sort of pipetting error that can occur quite easily in a complicated experiment with each observation derived from a different combination of concentrations is much less likely to occur in an experiment with a simple repetitive design. One is forced to the conclusion that even if major deviations from normality have been difficult to observe, this should give little support for believing that they do not occur.* Unfortunately, as discussed in Section 4.5, outliers do not have to be very gross or very numerous to undermine the theoretical basis for least-squares analysis.

What we shall be doing in this chapter will be searching for a *second-best method:* recognizing that in the real world our greed for obtaining 100 per cent efficiency in the ideal case may leave us with a disastrous loss of efficiency in a real case, we shall try to find a method that will work acceptably well in the ideal case, maybe not 100 per cent, but reasonably close, and will *also* work well in the less-than-ideal case.

Although there are two different problems here, shortage of weighting information and ignorance about the distribution, the ways of dealing with them are similar enough to warrant dealing with them together, and for hoping that a single method may overcome both. For the simplest case of the Michaelis–Menten equation, this is certainly possible: the median method (Section 5.5) proves to behave well both in the absence of good weights and when there are serious departures from normality. This approach becomes excessively cumbersome and computer intensive if one tries to adapt it to more complex problems, but fortunately other methods are available, as will be discussed.

Curiously, the statistical literature has manifested a great deal of interest in the distribution problem in recent years, especially since computer-intensive methods became cheap enough to be realistic, but almost none in the weighting problem. Possibly this is because the sort of large-scale engineering problems that attract the attention of statistical professionals offer enough data to make the choice of the right weights obvious or at least straightforward. Whatever the reason, enzymologists are largely on their own when it comes to choosing weights, but they can expect useful guidance from statistical texts in relation to robust regression—which in practice has often been limited to coping with outliers and other long-tailed departures from normality. Even in relation to outliers, robust methods

---

* A Popperian may object that this is an unscientific statement, if it is taken to suggest that we should continue to believe in major deviations from normality even if there is little or no evidence. But that is not the point: we are not concerned here whether non-normality constitutes 'the truth', but with whether we are willing to take the risks involved in believing the opposite. When you buy a fire insurance for a house, it is not because you believe it will catch on fire; on the contrary, you probably believe it will not, but the insurance company has persuaded you that the insurance is very cheap compared with the risk you take if you refuse it.

were developed in enzyme kinetics as long ago as in any other science: I was quite surprised to be told by a professor of statistics in the early 1980s that as far as he was aware the median method of fitting the Michaelis–Menten equation (Cornish-Bowden and Eisenthal 1974) was the only robust regresson method that had a significant level of usage in real experiments, in contrast to the many methods discussed in theoretical papers.

5.2   Median estimate of the slope of a straight line

For the simplest problem, that of estimating the population mean from a sample of values, the ordinary median is the archetype of a robust estimator. Even if the true distribution is normal it is 63.7 per cent efficient for an infinite sample and about 70 per cent efficient for samples of 5 to 20 observations (Section 4.5)—by no means an intolerable price to pay for almost total protection against the grossest of outliers. Nonetheless, we may hope to do better, because the median offers more protection against outliers than is really needed, as gross outliers are easy enough to recognize. For the moment, however, let us be satisfied with the level of efficiency offered by the median and ask whether the idea can be generalized to problems of model fitting, starting with the slope of a straight line, as the archetype of such problems.

It turns out that there are two different ways of generalizing the median to multi-parameter problems (Haldane 1948); this is because there are two ways of defining the ordinary median that give exactly the same results even though they are conceptually different. The familiar definition of the median is what Haldane called the *arithmetic median:* it is the middle one of the ranked values if the number of values is odd, or the mean of the middle two if it is even. Although medians can be either weighted or unweighted, as illustrated in Fig. 5.1, weighting usually makes so little difference to the result (Bowley 1928) that it is rarely done.

The alternative definition, the *geometric median*, is the value that leads to a minimum value of the sum of absolute values of the deviations of the observed values from the median (Laplace 1798). The parallel here with the ordinary arithmetic mean, which as well as being the sum of weighted values divided by the sum of weights is also the least-squares estimator (Section 1.4, eqn 1.4), suggests that least-absolutes regression might provide a robust alternative to least-squares regression, providing the same advantages (and disadvantages) that the median has in comparison with the mean. To some degree this is true (Section 5.6) but for the moment we shall consider the generalization derived from the ordinary (arithmetic) definition of the median.

For any straight line $y = a + bx$, a unique slope b_{ij} can always be

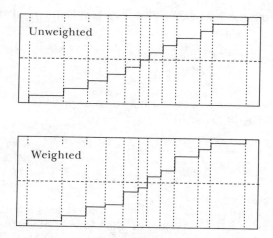

Fig. 5.1. *Unweighted and weighted medians.* The median is usually unweighted, and corresponds to the middle step when each step is of the same height. However, a weighted median is also possible, when each step has a height proportional to the weight assigned. As the most heavily weighted observations are more likely to occur in the middle of the sample, and *vice versa*, weighting normally has very little effect.

estimated from any pair of points (x_i, y_i) and (x_j, y_j) as long as $x_i \neq x_j$:

$$b_{ij} = \frac{y_i - y_j}{x_i - x_j}.$$

(5.1)

(If $x_i = x_j$ the denominator is zero. This does not mean that no slope is consistent with the observations, but that an infinite set of equally consistent slopes exists; however, the implication is still that no slope can be estimated from a duplicate pair.) For a set of n non-replicate observations, $n(n-1)/2$ such slopes can be determined, and the question then arises as to whether they can be combined in a single best estimate. It is fairly obvious that an ordinary unweighted mean will not be a good idea, because some of the estimates are likely to be quite bad, especially for neighbouring points (i.e. points for which $|x_i - x_j|$ is small). Nonetheless, one might hope to be able to compensate for this by giving more weight to values derived from more widely separated points, for example giving each estimate a weight w_{ij} defined as

$$w_{ij} = (x_i - x_j)^2,$$

(5.2)

and then defining the weighted mean $\tilde{b}$ as best estimate of the slope:

$$\tilde{b} = \frac{\sum\sum w_{ij}b_{ij}}{\sum\sum w_{ij}} = \frac{\sum\sum(x_i - x_j)(y_i - y_j)}{\sum\sum(x_i - x_j)^2}, \tag{5.3}$$

where each double summation means summation over all i from 1 to n and all j from 1 to n. Multiplying out, we have

$$\tilde{b} = \frac{\sum\sum(x_i y_i - x_i y_j - x_j y_i + x_j y_j)}{\sum\sum(x_i^2 - 2x_i x_j + x_j^2)} = \frac{2n\sum xy - 2\sum x\sum y}{2n\sum x^2 - 2(\sum x)^2}, \tag{5.4}$$

and comparison with eqn 1.43 shows that this is identical with the unweighted least-squares estimate of the slope. Calculating a weighted mean of the individual slope estimates does not, therefore, lead to a new way of estimating the slope, but to a different way of looking at a familiar way of estimating the slope. The importance of the result is to illustrate that median estimates can be defined in a very natural way so that they have the same relationship to least-squares estimates as the ordinary median has to the ordinary mean. Thus instead of taking the weighted mean of all b_{ij} values as the slope estimate, one takes the (unweighted) median:

$$b^* = \text{median}\left(\frac{y_i - y_j}{x_i - x_j}\right). \tag{5.5}$$

It might seem more logical to use a weighted median. However, as noted above, the median is not much affected by weighting (unlike the mean), and simulation of this method of estimating the slope (Cornish-Bowden, unpublished work) suggests that although weighting can give somewhat better results with some error distributions, it does not do so for all and can sometimes give worse results; in general it is difficult to avoid the conclusion that weighting a median is rarely worth the effort.

5.3   Confidence limits for median slope estimates

The median owes its robustness as an estimator to the fact that it depends on fewer and weaker assumptions than those of least squares. In contrast to the list of assumptions given in Section 4.7, the theory of the median is largely based on a single assumption, that any error is as likely to be negative as positive. This is just a weaker version of the last of the assumptions listed for least squares, that the observations are unbiased, i.e. their errors have zero mean, and to emphasize that it is a weaker version of this assumption we can state it as an assumption that the observations are *median unbiassed*. Because of the absence of any assumption about the distribution function, methods based on this idea are often called *distribution-free* methods. A term that overlaps in meaning, and is sometimes used as if it were synonymous, is *non-parametric;* this is not strictly equivalent, however, as it refers more to the absence of any use of the parameters that define the shape of the distribution curve.

For a set of observations of the same value, the assumption that they are median unbiassed leads to a very simple way of arriving at confidence limits for the true value. Even if one never has occasion to use this method, it fulfils a valuable didactic function, as it can be understood with no knowledge of distribution theory, indeed, with only as much understanding of probability theory as is needed for analysing coin-tossing experiments. It accordingly provides much insight into the meaning that statisticians apply to terms such as *confidence limit* and *likelihood* in more complex circumstances. Consider the set of five numbers (3.2, 5.6, 8.8, 1.2, 25.7), and suppose that they are median unbiassed observations of some unknown true value β. If β were less than 1, all the errors would be positive, with probability the same as for observing five heads in five tosses of an unbiassed coin, i.e. it is 0.5^5, or 0.03125; if β were between 1.2 and 3.2, we would have one negative error and four positive errors, with probability 5×0.5^5, or 0.15625; the complete set of possibilities is shown in Table 5.1.

As the probabilities are additive, we can calculate, for example, that the probability of having two or three positive errors is $0.3125 + 0.3125 = 0.625$, which gives us a way of defining 62.5 per cent confidence limits for β, the range from 3.2 to 8.8. This compares approximately with the

Table 5.1   Binomial probabilities

The table shows the calculation of the probability for each range of possible values of a quantity β, for which five observations yielded values 3.2, 5.6, 8.8, 1.2, and 25.7. Each of these observations is assumed to have a probability 0.5 of being greater than the true value. No other assumptions about their distributions are made.

Range for true value	Expression for probability	Probability
$\beta < 1.2$	$0.5^5 \times 5!/(5!0!)$	0.03125
$1.2 < \beta < 3.2$	$0.5^5 \times 5!/(4!1!)$	0.15625
$3.2 < \beta < 5.6$	$0.5^5 \times 5!/(3!2!)$	0.31250
$5.6 < \beta < 8.8$	$0.5^5 \times 5!/(2!3!)$	0.31250
$8.8 < \beta < 25.7$	$0.5^5 \times 5!/(1!4!)$	0.15625
$\beta > 25.7$	$0.5^5 \times 5!/(0!5!)$	0.03125
Total		1.00000

idealized interpretation of a standard error of the mean for a normally distributed sample, which corresponds in the limit of infinite sample size to a 68 per cent confidence region. With the median-unbiassed assumption we cannot choose the exact level of confidence; we can only choose among particular values that happen to be available.

Although the arguments are more complicated for a distribution-based calculation, the interpretation of a confidence limit is the same as here: given a particular assumption about the true parameter value, we can calculate the probability of observing the particular combination of errors that has been observed. In other words we are calculating the likelihood of the observations from an assumption about the truth. Much as we might like to do so, we are *not* calculating the probability of some particular hypothesis about the truth. The difference between the two can be seen by modifying the above example slightly. Suppose the second largest value were 5.6 instead of 8.8. It is obviously impossible for β to lie between 5.6 and 5.6, so for this particular set of observations it would be impossible to have two positive errors and three negative errors, and it would be absurd to assert that there is a 31.25 per cent probability that β lies in this particular 31.25 per cent confidence region. None of this alters, however, the calculation that in the universe of possible experiments that might be done in the same way, 31.25 per cent contain two positive errors and three negative errors.

The same sort of arguments can be used to generate joint confidence regions for models with two or more parameters (see Section 5.5, Fig. 5.3), but they are not very convenient in practice because the regions are irregular polygons that are not easily reduced to numbers that can be shown in a table, and they have been very little used. To obtain convenient confidence limits for the median estimate of a slope (and by extension to other parameters) we must add to the assumption that the observations are median unbiassed an assumption that all the errors come from the same distribution. (Although this is necessary for the argument below to be rigorous, it appears not to be necessary in practice, because computer simulations indicate that the calculated confidence regions are reasonably reliable even if the assumption that all the errors come from the same distribution cannot be true.)

If all the errors come from the same distribution, it follows that all rankings of these errors are equally likely, and hence that there is no correlation between the ranking of the errors and the ranking of the x values. How can we assess this? There are, in fact, two different correlation coefficients based on ranks that are in common use. The better known of these is Spearman's rank correlation coefficient, which is the same as the ordinary product–moment correlation coefficient with values replaced by ranks, but its statistical properties for small sample sizes are extremely complicated, and it does not provide a convenient basis for assessing the confidence of a median slope estimate. The other is Kendall's τ (Kendall 1970), which proves to be ideal for this purpose. It is defined as follows, for measuring the correlation between n pairs of values x and e:

$$\tau = \frac{\sum \sum f(e_i - e_j, x_i - x_j)}{n(n - 1)}, \tag{5.6}$$

where the double summation sign has the same meaning as in the previous section, i.e. $i = 1$ to n and $j = 1$ to n, and

$$f(e_i - e_j, x_i - x_j) = \begin{cases} 0 & i = j \\ 1 & i \neq j \quad \text{and} \quad \dfrac{e_i - e_j}{x_i - x_j} > 0 \\ -1 & i \neq j \quad \text{and} \quad \dfrac{e_i - e_j}{x_i - x_j} < 0. \end{cases} \tag{5.7}$$

(For simplicity I shall ignore the possibilities that $e_i = e_j$ or $x_i = x_j$ when $i \neq j$, but these can be taken into account in a full analysis.) If $e = y - a - b^{**}x$, i.e. the difference between observed and calculated y values for a slope b^{**}, then eqn 5.7 may be written as follows:

$$f(e_i - e_j, x_i - x_j) = \begin{cases} 0 & i = j \\ 1 & i \neq j \quad \text{and} \quad b_{ij} > b^{**} \\ -1 & i \neq j \quad \text{and} \quad b_{ij} < b^{**}. \end{cases} \tag{5.8}$$

where b_{ij} is defined as in eqn 5.1, i.e. $b_{ij} = (y_i - y_j)/(x_i - x_j)$. It is then clear that $\tau = 0$ if $b^{**} = b^*$ as defined by eqn 5.5, i.e. the median slope estimate is the estimate that gives zero correlation if this is measured using Kendall's τ.

This result is of practical as well as theoretical interest, because it provides a way introduced by Sen (1968) for defining confidence limits for the median slope b^*. The distribution of Kendall's τ can be calculated for the hypothesis that the true value is zero. This can be done exactly, but it requires each case (not only each value of n, but each grouping of n observations into replicates) to be dealt with as a separate special case; it is more convenient to use the fact that the distribution is almost normal even for small n and is virtually exact for $n > 20$. This known distribution can then be used to set confidence limits for the true slope β.

5.4   Relationship between least-squares and median estimates

We have seen in Section 5.2 that the least-squares and median estimates of the slope can be regarded as different ways of averaging the slope estimates given by all pairs of non-replicate observations. However, the particular weighting needed to arrive at eqn 5.4 may have appeared rather arbitrary, designed with a particular aim in view. The demonstration that the median estimation corresponds to setting Kendall's τ to zero shows a more fundamental relationship, as both kinds of estimate correspond to setting to zero a correlation coefficient between independent variable and residuals.

The most familiar correlation coefficient is the product–moment coefficient ρ, defined for two variates x and e as follows:

$$\rho = \frac{\sum (x - \bar{x})(e - \bar{e})}{\left[\sum (x - \bar{x})^2 \sum (e - \bar{e})^2 \right]^{1/2}}. \tag{5.9}$$

For this to be zero it is sufficient for the numerator to be zero, so the denominator can be ignored. Thus if e is written as $y - a - bx$ and $\bar{e}$ as $\bar{y} - a - b\bar{x}$, $\rho = 0$ if

$$\sum (x - \bar{x})(y - a - bx - \bar{y} + a + b\bar{x}) = 0. \tag{5.10}$$

This may be solved to give the value of b:

$$b = \frac{n \sum xy - \sum x \sum y}{n \sum x^2 - \left(\sum x \right)^2}. \tag{5.11}$$

This is identical with the least-squares estimate of the slope if all weights are set to unity. It follows, then, that the median estimate of the slope of a straight line has exactly the same relationship to the unweighted least-squares estimate as Kendall's τ has to the product–moment correlation coefficient.

5.5   Median estimates of Michaelis–Menten parameters

The approach of Sections 5.2 and 5.3 can be applied to the Michaelis–Menten equation $v = Va/(K_m + a)$ by defining $1/V^*$ as the median slope estimate for two-point plots of a/v against a,

$$1/V^* = \text{median} \left[\frac{(a_i/v_i) - (a_j/v_j)}{a_i - a_j} \right], \tag{5.12}$$

and K_m^*/V^* as the median slope estimate for two-point plots of $1/v$ against $1/a$,

$$K_m^*/V^* = \text{median} \left[\frac{(1/v_i) - (1/v_j)}{(1/a_i) - (1/a_j)} \right]. \tag{5.13}$$

V^* and K_m^* can then be calculated from these.

This way of calculating median estimates of Michaelis–Menten parameters (Cornish-Bowden and Eisenthal 1978) is somewhat different from the original proposal (Cornish-Bowden and Eisenthal 1974) to calculate V^* and K_m^* directly from the following expressions:

$$V^* \approx \text{median}\left[\frac{a_i - a_j}{(a_i/v_i) - (a_j/v_j)} \right], \qquad (5.14)$$

$$K_m^* \approx \text{median}\left[\frac{v_j - v_i}{(v_i/a_i) - (v_j/a_j)} \right], \qquad (5.15)$$

in which the approximate equality sign $\approx$ emphasizes that these are *not* equivalent to the indirect calculations using eqns 5.12 and 5.13. A trap that can easily escape notice is that the median of a set of values is not the same as the reciprocal of the median of their reciprocals, because of different ways in which numbers with opposite signs are ranked (there is an additional difference if the number of values is even, but this is usually trivial): median$(-1, 1, 2) = 1$, but $1/\text{median}(-1/1, 1/1, 1/2) = 2$.

This argument shows that calculating V^* and K_m^* directly by means of eqns 5.14 and 5.15 is not the same as calculating them indirectly by means of eqns 5.12 and 5.13, but it does not show which approach is to be preferred. Analogy with least-squares regression, where $1/V$ and K_m/V emerge from the analysis as the natural pair of parameters (Section 2.1) for fitting the Michaelis–Menten equation suggests that eqns 5.12 and 5.13 will make the better choice. As the problem arises from the large differences in how negative parameter estimates are handled, it can be resolved by examining how such negative estimates arise and how they ought to be interpreted. It is also worth noting that as v_i and v_j occur in the denominators of eqns 5.14 and 5.15, these denominators can be zero even if the observations are not replicates.

Figure 5.2 illustrates the principal possibilities. In the ideal case (Fig. 5.2(a)), the errors in the two points are small enough that a rectangular hyperbola defined by positive values of V_{ij} and $K_{m,ij}$ can be drawn through them. However, if the smaller rate has a large enough negative error and the larger rate a large enough positive error to define an upward rather than a downward curvature (Fig. 5.2(b)), the corresponding V_{ij} and $K_{m,ij}$ are both negative. Finally, if $a_i > a_j$ whereas $v_i < v_j$ (Fig. 5.2(c)), $K_{m,ij}$ is negative but V_{ij} is positive. (The fourth case, positive $K_{m,ij}$ with negative V_{ij}, is impossible if v_i and v_j are positive.)

If a series of observations consistently suggests one or other of these anomalies, the obvious interpretation is that the data do not fit the

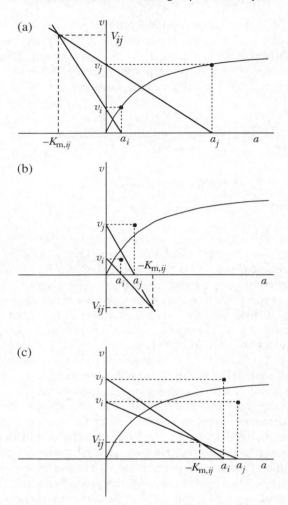

Fig. 5.2. *Negative estimates of Michaelis–Menten parameters.* (a) Ideally a rectangular hyperbola corresponding to positive values of both V and K_m can be drawn through any pair of points. (b) However, at low substrate concentrations experimental error may cause the pair of points to suggest upward curvature, which requires negative values of both V and K_m. (c) At high substrate concentrations the pair of points may suggest that v decreases with increasing a, which requires positive V and negative K_m.

Michaelis–Menten equation. Thus, a series of points like those in Fig. 5.2(b) suggests a sigmoid dependence of rate on concentration; a series of points like those of Fig. 5.2(c) suggests substrate inhibition. However, if the anomalies are rare it is more reasonable to attribute them to experimental error, and in this case the question arises as to how the values of

V_{ij} and $K_{m,ij}$ ought to be interpreted. The errors are exaggerated in Fig. 5.2 to make it possible to illustrate the effects on a convenient scale and with intermediate values of a_i and a_j. In reality the case of Fig. 5.2(b) is most likely to occur when the true values of v_i/a_i and v_j/a_j are almost equal, i.e. the points are close to the origin, so that a_i and a_j are both small compared with K_m, and v_i and v_j are both small compared with V. It follows then that negative V_{ij} and $K_{m,ij}$ ought to be interpreted as 'beyond infinity', i.e. as very large values in the rankings, not as very small ones.

The case of Fig. 5.2(c) is most likely to occur when the true values of v_i and v_j are almost equal, i.e. the enzyme is approaching saturation, so that a_i and a_j are both large compared with K_m, and v_i and v_j are similar in magnitude to V. Treating negative $K_{m,ij}$ at face value while determining the median is now quite reasonable: it just means that the particular pair of observations is indicating that K_m is small compared with a_i and a_j.

It follows, then, that the way negative $K_{m,ij}$ is interpreted varies with whether V_{ij} is also negative: if so, both $K_{m,ij}$ and V_{ij} should be interpreted as beyond infinity, i.e. as very large, but if V_{ij} is positive both estimates may be taken at face value. As these rules may seem complicated, and even arbitrary, one may ask whether there is a more straightforward way of arriving at a similar result, and in fact using eqns 5.9 and 5.10, i.e. treating $1/V$ and K_m/V as the primary parameters, provides this. If these equations are examined it is evident that no infinities can arise unless $a_i = a_j$; thus the only special rule to be followed in determining the medians is the intuitively obvious one that values derived from replicate observations must be excluded.

The use of Kendall's τ to set confidence limits on the slope described in the previous section can be applied to Michaelis–Menten parameters if one is willing to treat both $1/V$ and K_m/V as slopes of straight lines (Cornish-Bowden *et al.* 1978). It is impossible, however, for all differences between observed and calculated a/v values to be identically distributed, and simultaneously for all differences between observed and calculated $1/v$ values to be identically distributed. At best, therefore, the underlying assumption can be true for one parameter or the other, but it cannot be true for both. Nonetheless, tests with simulated data (Cornish-Bowden *et al.* 1978) indicate that in practice the confidence limits calculated with Kendall's τ contain the true values with about the expected frequency.

A joint confidence region for $1/V$ and K_m/V based on the median-unbiased assumption only (i.e. without assuming that the different deviations are identically distributed) can be set up as illustrated in Fig. 5.3. As 16 different sequences of signs are possible for four observations, and as each is equally likely *a priori* according to the median-unbiased assumption, each sign sequence constitutes a 6.25 per cent confidence region. Many of these regions extend to infinity; the finite regions are those with at least three runs of signs, i.e. regions 2 $(- - + -)$, 4 $(- + - -)$, 5

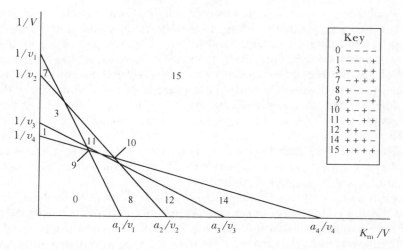

Fig. 5.3. *Confidence regions for the direct linear plot.* A set of (a, v) values is plotted in $(K_m/V, 1/V)$ space such that each observation is represented by a straight line with intercepts a/v and $1/v$ on the K_m/V and $1/V$ axes respectively. Each line corresponds to a $(K_m/V, 1/V)$ pair that satisfies the corresponding observation exactly, and divides the entire space into two regions that differ according to the sign of the deviation of the line from the value of v calculated for any point in the region, and each of the labelled regions corresponds to a different combinations of signs for the four observations. For example, if the true values of K_m/V and $1/V$ correspond to a point in the region labelled 9, the errors in the four v values correspond to the sign combination $+ - - +$. The labels are the decimal equivalents of the binary numbers produced by writing $-$ as 0 and $+$ as 1, e.g. binary 1001 is decimal 9.

$(- + - +)$, 6 $(- + + -)$, 9 $(+ - - +)$, 10 $(+ - + -)$, 11 $(+ - + +)$, and 13 $(+ + - +)$. As there are eight of these, the central finite region of the plot constitutes a 50 per cent joint confidence region.

This example illustrates clearly what a confidence region is and what it is not. It is clearly not correct to say, for example, that there is 0.0625 probability that the true values in this particular experiment lie in region 6 of Fig. 5.3: there is no region 6, because the particular sign sequence $- + + -$ is impossible for this set of data. Clearly the probability is zero that the true values are in region 6 for this experiment. What one must say is that the true values lie in region 6 in 6.25 per cent of the universe of experiments that might be done in the same way. By extension, we cannot say that there is a 50 per cent probability that the true values lie in this particular central finite region; we must say that in 50 per cent of experiments done in the same way the true values lie in a central region defined in the same way.

I have used an experiment with only four observations in order to produce regions that are easily distinguishable in the figure, for clearer presentation of the ideas. When there are more observations the region with three or more runs of signs corresponds to higher levels of confidence, and if there are 12 or more observations the region corresponding to four or more runs corresponds to a region of greater than 95 per cent confidence (for full details, see Cornish-Bowden and Eisenthal 1974). Daniels (1954) suggested a similar way of defining confidence regions in straight-line regression, based not on runs of signs but on the number of lines that have to be crossed to reach any finite region from the infinite peripheral regions.

5.6   Least absolutes fitting

As noted in Section 5.2, the median can be defined not only in the conventional way as the middle value of the ranked set, but also as the value from which the sum of absolute deviations is a minimum. For a univariate sample the definitions are exactly equivalent, but they become different when they are generalized. The ordinary median estimates of Michaelis–Menten parameters (Section 5.5) constitute one kind of generalization; least-absolutes estimates provide another. For any model for which the least-squares estimates are obtained by minimizing a sum of squares SS defined as

$$SS = \sum w(v - \hat{v})^2, \tag{5.16}$$

corresponding least-absolutes estimates can be defined as those that minimize the sum of absolute deviations SA defined as

$$SA = \sum w^{1/2}|v - \hat{v}|, \tag{5.17}$$

in which w is raised to the power $\frac{1}{2}$ so that the same w applies to both expressions.

Simulations with the Michaelis–Menten equation (Cornish-Bowden and Eisenthal 1974) indicate that minimizing SA provides at least as much protection against outliers as taking median estimates, maybe somewhat more, but that it requires correct assumptions about weights to work well. If information about the weighting is not available, or if incorrect assumptions are made, least-absolutes estimates tend to be intermediate in quality between least-squares estimates and median estimates. This intermediate behaviour follows naturally from the powers of w that appear in the different methods: using a wrong value of w is more serious than using a wrong value of $w^{1/2}$, which is itself more serious than not using w at all.

It follows that if there is some independent information about weighting, and if outlier protection is considered necessary, minimizing SA may provide an attractive option, especially as it is readily generalized to multi-parameter problems, whereas median estimation is not. It presents a different kind of difficulty, however, in that the actual methods used to minimize SA are in general fundamentally different from those that are used to minimize SS. This is because most of the latter use, either explicitly or implicitly, the fact that SS is a smooth function of the parameter values with defined partial derivatives with respect to the parameter values at all points. (Consider, for example, how difficult it would have been to minimize the function defined by eqn 1.38 if one had not been allowed to use eqns 1.39 and 1.40 or anything equivalent to them.) Equation 5.14 does not define a smooth function, and the partial derivative of SA with respect to any parameter value shows an abrupt 'saltus', or jump, from one value to another when the parameter passes through a value at which any $v - \hat{v}$ is zero. This means that one cannot use the local behaviour of SA in some region of parameter space as a guide to how it will behave in some other region; in other words one cannot predict where the minimum ought to be.

Fortunately, however, alternative methods exist which, although quite different from those used for least squares, are not much more laborious. The essential idea depends on the fact that the least-absolutes fit to any linear equation with p parameters must fit p observations exactly (Fisher 1961). To understand why this is so, consider the behaviour of SA as a function of the slope b of a straight line:

$$SA = \sum | y - a - bx |. \tag{5.18}$$

For simplicity $w^{1/2}$ is omitted from the definition of SA, though the argument can readily accommodate weighting, and all the x values will be assumed to be positive, though again this is not essential to the argument. Regardless of the value of a, we can initially make b so large that all the individual $y - a - bx$ values are negative. If now we decrease b, every one of these negative values must decrease in absolute magnitude, so SA must decrease, and this trend must continue monotonically as long as it remains true that all individual $y - a - bx$ values are negative. Every time we cross a value of b that fits any observation exactly, however, the corresponding $y - a - bx$ value changes sign, so that further increases in b generate one or more increasing contributions to SA. As each $y - a - bx$ changes sign only once when we scan through the whole range of b values, it is evident that the plot of SA against b must consist of a series of straight lines with abrupt changes of slope, always in the same direction at each b value for which an observation is exactly satisfied. Such a plot must either give a unique minimum SA value at one of the abrupt changes in slope, or it must have a flat minimum along a line of zero slope connecting two such points. In either case the value of b that minimizes SA must satisfy at least one observation exactly. If this is true at an arbitrary value of a, it must be true at the particular value of a that minimizes SA. The same argument applies to a at any value of b and, more generally, for a linear model with p parameters it follows that the minimum solution must satisfy p observations exactly.

This argument also excludes the possibility of multiple minima in a linear model, because whichever parameter is chosen to vary there must be either a unique lowest point or a horizontal line joining two points. This has the important consequence that to find the minimum SA it is not necessary to sample all combinations of p parameters (which is highly time consuming if p is greater than 2 and there are many observations). Instead, it is sufficient to proceed as follows.

1. Choose $(p-1)$ observations at random, and hold these choices constant while systematically checking all possibilities for the pth observation.

2. For each set of p observations calculate the parameter values to satisfy them exactly and calculate SA with these parameters.

3. When the best choice for the pth observation is found, vary the choice of a different one of the original $(p-1)$ observations in the same way to find the best choice for that observation.

4. Repeat this for every observation in turn.

5. When all observations have been varied, start again from the beginning, because the choice found for the first observation varied may no longer be the best after the others have been varied.

6. Continue until it is no longer possible to find a new minimum, i.e. until any change from the best solution found leads to a worse solution.

In this way an impractical and time-consuming search in p dimensions is converted into a one-dimensional search, because it is never necessary to vary more than one choice of observation at a time.

Application of similar arguments to a non-linear model leads to the difficulty that the lines joining the points of abrupt changes in slope are not straight but curved, so it is not impossible for the minimum SA to occur between two such points. However, the models of interest in steady-state enzyme kinetics generally behave sufficiently well for this objection to be more hypothetical than real, and even if the true minimum does not exactly correspond to a point satisfying p observations, it will be close enough for practical purposes.

6

Robust regression

6.1 Recognizing and dealing with outliers

All published proposals for rejection criteria, based on any kind of mathematical reasoning, from Pierce's (1852) onwards, have an unexplained starting point or objective, presented as if it were the only obvious one and in fact utterly obscure.

Anscombe 1960

The most useful information is not infrequently found to reside in the apparent wild shots themselves.

Tukey 1962

The median and its generalizations to model-fitting problems provides a straightforward way of dealing with outliers: it allows them to remain in the calculation on an equal basis with other observations until the end, but it does not allow them to distort the results. However, because using median estimation requires abandoning least-squares estimation, as it is not easy to imagine a smooth transition from one to the other that might allow compromise solutions, it can easily appear rather an extreme way of dealing with the short-comings of least squares. To achieve adequate efficiency when least squares breaks down completely one has to tolerate rather low efficiency in circumstances where least squares performs well.

A further difficulty is that the approach outlined in Sections 5.2–5.5 becomes computationally very cumbersome when the number of parameters increases, especially if account is taken of the tendency of experimenters to increase the number of observations when they increase the complexity of the models they wish to fit. Thus considering all possible combinations of two out of ten observations requires 45 sets of two simultaneous equations to be solved; considering all possible combinations of four out of 40 observations requires 91 390 sets of four simultaneous equations to be solved, much more than a 2000-fold increase in computational effort for a four-fold increase in experimentation. A minor additional difficulty (minor because it presents no problem for a properly written computer program) is that whereas it is easy to recognize the combinations that lead to insoluble (singular) simultaneous equations in the two-parameter case, it is less easy to recognize them by eye when there are three or more parameters: not only replicate observations but also observations consisting of linear combinations of the same values of the independent variables must be excluded.

An alternative to median estimation is to stay with least squares but to add methods for recognizing and dealing with outliers so that they cannot grossly distort the results. Traditionally, outlier handling was an all-or-nothing approach: either an observation was an outlier, in which case it was removed completely from the calculation, or it was not, in which case it received no special treatment. This is still the approach if the initial suspicion that an observation might be an outlier is followed by *independent* confirmation that it was in some important way different from the others. Suppose, for example, that after fitting a set of observations to the Michaelis–Menten equation one point deviates by 25 per cent from the fitted line whereas all the other deviate by less than 2 per cent. Is the point an outlier? From the deviation alone one cannot say with confidence, but suppose that re-examination of the laboratory notebook reveals that the suspect observation was made a day after all the others, and that it was done with a substrate from a different supplier because the original stock was exhausted. Clearly in such a case there are good grounds other than the magnitude of the deviation for rejecting the observation.

This is, however, an extreme example. Often there will be nothing apart from a large deviation to distinguish a suspect point from the others. In such cases it is unwise (or even dishonest, in published results) to discard the observation, because the large deviation may contain real information, for example that the true model is more complicated than we thought (we should not forget the advice of Tukey (1962) given at the beginning of this section). It is also too extreme, because there may be gradations of reliability between totally reliable and totally unreliable. Modern tendency, therefore, is not to reject outliers but to assign decreased weight to them, in the limit zero weight.

Now it may seem that giving zero weight to an observation is exactly the same as rejecting it, but there is an important difference. A zero-weighted observation always has the possibility to return during an iterative calculation, if the iterative trend is such that it appears less deviant in the later iterations than it did at the beginning. Even an observation that remains with zero weight at the end of the calculation is different from a rejected observation, because it appears (or should appear) in the published results, whether in graphs or tables, so that it is available to readers who may wish to interpret the data in a different way from that of the authors.

6.2   Biweight regression

The *biweight* (or *bi-square weighting*) method of Mosteller and Tukey (1977) is one of the most satisfactory ways of implementing the ideas of

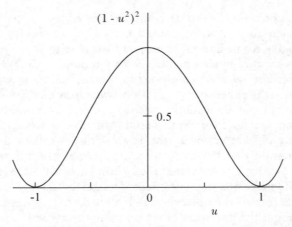

Fig. 6.1. *Basis for the biweight method.* The function $(1 - u^2)^2$ of u has the desirable property of descending smoothly to zero as $|u|$ approaches 1, but the undesirable property of increasing as $|u|$ exceeds 1.

the previous section. It is a modification of least squares in which all observations are initially given full weight, but the weights are modified in subsequent iterations as a function of their deviations from the fitted equation. The starting point for discussion is that as the method of least squares has maximum efficiency for normally distributed data, a good method should behave as nearly as possible like least squares for observations with small to moderate deviations, but should assign small or zero weight to observations with large deviations.

Thus, if we have some standardized measure of deviation u, we need to calculate weights from a suitable bell-shaped function of u that has a broad maximum when $|u|$ is small but decreases smoothly to zero as $|u|$ increases. As illustrated in Fig. 6.1, the function $w(u^2 - 1)^2$ has just these properties when $|u| < 1$, although it has the undesirable property of increasing again when $|u| > 1$. We can thus define satisfactory robust weights W as follows:

$$W = \begin{cases} w(u^2 - 1)^2 & \text{if } |u| \le 1 \\ 0 & \text{if } |u| > 1 \end{cases}, \qquad (6.1)$$

in which w is the ordinary weight calculated from the heteroscedasticity of the observations. There are three points that have to be decided before these can be used, however: the definition of the deviation u has to be specified; the heteroscedasticity of the observations needs to be known so that w can be calculated; as no deviation can be measured without a model, any calculation needs to be iterative. For the moment we shall ignore the second of these, i.e. we shall assume we have some way of defining w, but this will be discussed in Section 6.3.

The starting point for defining u is obviously the weighted difference between the observed and calculated values of the dependent variable, which we shall assume to be a rate v, i.e. $w^{1/2}(v - \hat{v})$. However, as this has the units of v it needs to be defined in a dimensionless form, i.e it needs to be suitably scaled. Given that W becomes zero when $|u|$ exceeds 1, it follows that the scaling is decided by the magnitude of deviation that we consider so large that the observation should contribute nothing to the final estimate. This magnitude is most naturally defined as some multiple c of a 'typical' value S of $w^{1/2}(v - \hat{v})$ determined from the data. There are various ways of defining S, but a convenient one is as the median absolute value of $w^{1/2}(v - \hat{v})$, which is reasonably resistant to the effects of outliers. The factor c can be defined to reflect the degree of robustness desired. It is it made very large all observations give small values of $|u|$ and the method becomes indistinguishable from least squares. If it is made small many observations have substantially decreased weight, but one should be cautious about making it too small (say less than 3) because then most observations will have very small weight and those that happen to have the smallest deviations after the first iteration will tend to dominate the final result. Mosteller and Tukey suggest that values in the range 6–9 are best.

In summary, therefore, the deviations u needed for calculating weights from eqn 6.1 are defined as follows:

$$u = w^{1/2}(v - \hat{v})/cS \qquad\qquad (6.2)$$

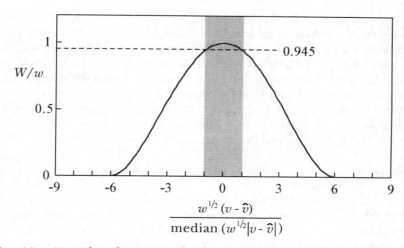

Fig. 6.2. *Biweight adjustment.* The factor W/w needed for converting the ordinary heteroscedastic weights w into robust weights is calculated as described in the text. For deviations less than the median deviation the biweight adjustment has only a trivial effect, as shown by the shading, but very large deviations result in zero weight.

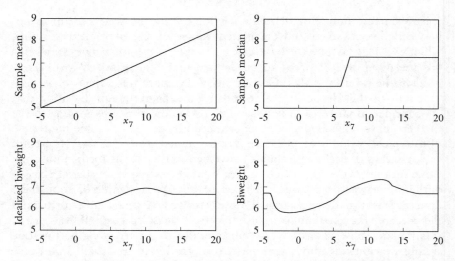

Fig. 6.3. *Sensitivity curves for estimators of the population mean.* The four curves show how the sample mean, the sample median, and the biweight vary with the value of x_7 for the seven numbers $(9.3, 6.0, 3.1, 5.7, 8.5, 7.3, x_7)$. The curve for the idealized biweight shows how we would like the biweight to behave; the curve at bottom right shows how it actually behaves.

where c can be set at will but is typically 6 or 9, and

$$S = \text{median}(w^{1/2}|v - \hat{v}|). \tag{6.3}$$

Figure 6.2 shows the sort of weights produced by this definition. Note that half of the observations lie within the shaded central area, because of the way in which S is defined, and consequently have weights at least 94.5 per cent of their 'normal' weights w. A substantial proportion of the remainder (the exact proportion will depend on the distribution) have values of $6|u|$ less then 3.25, and consequently have weights of at least 50 per cent of w. Thus for the great majority of observations the biweight method is not very different from least squares, but the small number of exceptions make it much more resistant than least squares to the effects of outliers.

Figure 6.3 shows a set of *sensitivity curves* showing how various methods of estimating the population mean from a set of seven numbers depend on one particular value of the set as it varies from -5 to $+20$. The sample mean shows a simple straight line dependence that continues indefinitely in both directions, i.e. it offers no protection at all against large errors in x_7. The sample median offers a high degree of protection against outliers, but its behaviour is still far from ideal: over most of the range it does not vary at all with x_7, and the small range in which it does vary is arbitrarily dependent on the spacing of the other observations in the middle of their range.

The curve labelled 'idealized biweight' is the sort of curve sometimes shown in textbooks describing the biweight method: this is how we would *like* the biweight to behave. It mimics the sample mean in the middle of the range; but tails away smoothly and symmetrically from it as the deviation increases. At large deviations it moves in the opposite direction from the observation, because as the deviation increases in this region our confidence in the meaningfulness of the observation decreases rapidly, until at very high deviations we have no confidence in it at all and take the estimate as the sample mean of the remaining observations. This curve was, however, calculated in an unrealistic way, treating the other six observations as 'definitely not outliers' with weights constant at 1 and scale factor S constant at 1.4. In fact, of course, we would have no basis for singling out a single observation for special treatment, especially if its value were within the range of the others, and so we cannot treat any of the weights as constants while computing the sensitivity curve. All the weights vary, as does S, because the value of x_7 contributes to its value.

For these reasons the true sensitivity curve for the biweight, shown in the bottom-right panel of Fig. 6.3, is less tidy than the idealized curve: it is asymmetrical; it shows abrupt changes in slope reflecting abrupt changes in the value of S; it varies over a wider range of x_7 values than the ideal curve does, and it leads to a wider range of final values. Nonetheless, qualitatively its behaviour is very satisfactory: it offers great outlier protection, and it is much smoother and less arbitrary than the median.

The biweight is now probably the method of choice for dealing with outliers, and any serious modern program for data analysis ought to allow for its use. However, it addresses only one of the two difficulties that arise in traditional least-squares analysis, as it does nothing about lack of knowledge of the underlying heteroscedasticity of the observations, because it assumes that the ordinary weights w that need to be inserted in eqn 6.1 are known (or worse, it just replaces them by unit weights). The next section will address this deficiency.

6.3 Assessing heteroscedasticity

To this point we have assumed that the original observations that are to be analysed either have uniform variance or they have uniform coefficient of variation. This is admittedly an advance on many elementary treatments of regression, but it leaves a great deal of unanswered questions: why should nature be limited to just these possibilities? How do we recognize if it is not so limited? What can we do about it? In any case, although homoscedasticity is the same as uniform variance, heteroscedasticity is much broader than meaning just a uniform coefficient of variation. In the real world heteroscedasticity may imply quite complex variations in variance, with the variance a function of the true value of the observation and of the independent variables and other variables such as time, background temperature, etc. that are not explicitly included in the model at all.

All of this can in principle be taken into account if there is a very large number of observations and if the experimenter is willing to go to a great deal of trouble to analyse how the variance behaves. But this is rarely practicable in enzyme kinetics, and even more rarely done. The problem that we shall address in this section is thus a simpler one: is it possible for a computer program (without necessary human intervention) to make a reasonably reliable assessment of the heteroscedasticity of a *small* set of observations? When I first studied this problem in the 1970s my expectation was that it would not be possible unless 'small' was taken to mean 30 or more, because one needs at least that sort of number of observations to obtain a reasonably clear idea of the proper weight from residual plots. Fortunately, however, this expectation proved to be unduly pessimistic: computer simulations indicate that the approach to be described yields much better parameter estimates in general than can be obtained by just guessing the weights, even for experiments with as few as ten observations.

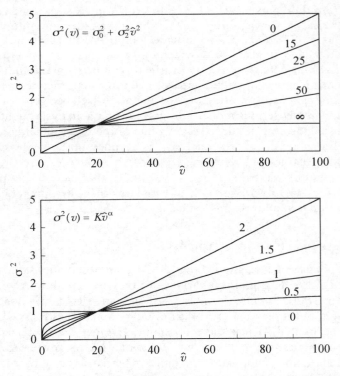

Fig. 6.4. *Two hypotheses for intermediate degrees of heteroscedasticity.* In the upper panel the curves are labelled with values of σ_0^2/σ_2^2; in the lower panel they are labelled with values of α. In all cases the additional parameter is set to give a variance of 1 when $\hat{v} = 20$.

Even if we limit the possible ways of defining heteroscedasticity to functions intermediate between uniform variance and uniform coefficient of variation, there are many ways of defining such functions, but I shall consider just two of them here. One is to suppose that the final variance is the sum of a constant and a term proportional to $\hat{v}^2$ (as always, we use the calculated value as a substitute for the ideal but unknown true value):

$$\sigma^2(v) = \sigma_0^2 + \sigma_2^2 \hat{v}^2. \tag{6.4}$$

The other is to suppose that it is directly proportional to some power of $\hat{v}$:

$$\sigma^2(v) = K\hat{v}^\alpha. \tag{6.5}$$

These are by no means equivalent, as may be seen from the representative curves plotted in Fig. 6.4. They differ especially in the predictions they make at extreme values: eqn 6.4 suggests that at high values of $\hat{v}$ the behaviour approximates a constant coefficient of variation, but that at low values there is a hard core of uncertainty that cannot be improved on; eqn 6.5, on the other hand, suggests that a zero can be measured exactly, but that at high values of $\hat{v}$ the behaviour slowly tends towards uniform variance. Thus if either of eqns 6.4 and 6.5 were the true function, the other equation could be considered to combine the worst attributes of uniform variance and uniform coefficient of variation, and vice versa. Despite this, the weights produced by the two equations, and hence the parameter values that they lead to, do not differ from one another as much as might be expected, as long as the range of v values is not very great. Figure 6.5 illustrates that despite the very important deviations at the extremes, the curves defined by eqns 6.4 and 6.5 may differ by only a few per cent over a considerable range.

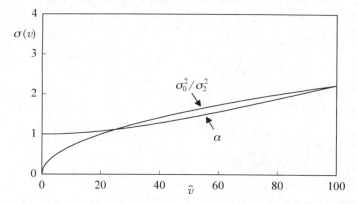

Fig. 6.5. *Similarity of weighting hypotheses.* Despite very important disagreement at the extremes, calculating the variance in terms of σ_0^2/σ_2^2 and α gives essentially the same results over a wide intermediate range.

We must now consider how to apply these equations in practice, given that we have no a priori information about the values of their parameters, and even we have enough observations to make estimating them from residual plots a realistic option, we should anyway prefer an automatic (programmable) approach that does not require necessary human intervention. The approach described here (Cornish-Bowden and Endrenyi 1981) uses eqn 6.4 as an example, but the treatment of eqn 6.5 is closely analogous.

For any variable, if there is some way of obtaining both observed and calculated values, v and $\hat{v}$ respectively, then for any one observation the value of $e^2 = (v - \hat{v})^2$ is an estimate of $\sigma^2(v)$, a bad estimate, to be sure, but better than no estimate at all. It follows then that eqn 6.4 defines a straight-line dependence of e^2 on $\hat{v}^2$, with intercept σ_0^2 and slope σ_2^2:

$$e^2 = (v - \hat{v})^2 \approx \sigma_0^2 + \sigma_2^2 \hat{v}^2. \qquad (6.6)$$

As this expression certainly has large errors of unknown distribution it would be rash to try to treat fitting it as an ordinary problem in straight-line regression; indeed, experience suggests that it is virtually impossible to obtain meaningful estimates of both parameters from small data sets. Fortunately, however, we do not require both parameter values for calculating weights: as long as we can estimate the *sigma ratio* σ_0^2/σ_2^2 (which has the same dimensions as v^2) we can calculate weights, because the second parameter is just a scale factor that can be set at any value that generates a convenient range of numbers for the weights.

Provided that reasonable rules are set up to deal with absurd estimates, σ_0^2/σ_2^2 can be estimated with acceptable accuracy by a method similar to the median method for fitting the Michaelis–Menten equation. For any pair of non-replicate observations i and j, an estimate $(\sigma_0^2/\sigma_2^2)_{ij}$ can be calculated as follows:

$$\left(\frac{\sigma_0^2}{\sigma_2^2}\right)_{ij} = \begin{cases} +\infty & \text{if } \dfrac{e_j^2 - e_i^2}{\hat{v}_j^2 - \hat{v}_i^2} < 0 \quad \text{(negative slope)} \\[2ex] 0 & \text{if } \dfrac{e_i^2 \hat{v}_j^2 - e_j^2 \hat{v}_i^2}{\hat{v}_j^2 - \hat{v}_i^2} < 0 \quad \text{(negative intercept)} \\[2ex] \dfrac{e_j^2 - e_i^2}{e_j^2 \hat{v}_i^2 - e_i^2 \hat{v}_j^2} & \text{otherwise} \end{cases}$$

$$(6.7)$$

and then σ_0^2/σ_2^2 can be estimated as the median of the individual values. The two special cases play a far more central role in this method than the special cases discussed in Section 5.5 for calculating median estimates of Michaelis–Menten parameters, when physically meaningless values typically occur sufficiently rarely that it is no great disaster if they are not correctly dealt with. When applying eqn 6.7, however, it can easily happen that more than half the estimates are 0 or $+\infty$ (so that it is not really

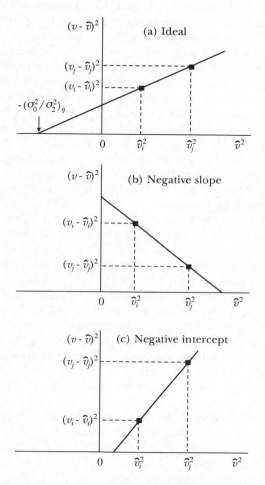

Fig. 6.6. *Interpretation of two-point residual plots.* (a) Ideally the straight line drawn through two points on a plot of $(v - \hat{v})^2$ against $\hat{v}^2$ has positive slope and ordinate intercept, and in this case the abscissa intercept is accepted as an estimate of $-\sigma_0^2/\sigma_2^2$. (b) However, the line may have a negative slope, suggesting negative variance at high $\hat{v}^2$. As this is considered impossible, such a pair of points is considered to indicate constant variance, i.e. that σ_0^2/σ_2^2 is very large. (c) Alternatively, the line may suggest negative variance at low $\hat{v}^2$, and as this is likewise considered impossible it is taken to indicate variance proportional to $\hat{v}^2$, i.e. σ_0^2/σ_2^2 is zero.

correct to regard them as special cases at all), and it is important to understand why they arise.

In the ideal case (Fig. 6.6(a)), a two-point plot of $(v - \hat{v})^2$ against $\hat{v}^2$ yields a straight line with positive slope and positive intercept on the ordinate, and it is clear that $-(\sigma_0^2/\sigma_2^2)_{ij}$ is the intercept on the abscissa. However, if the slope is negative (Fig. 6.6(b)), the naively drawn straight line implies that the variance becomes negative at large $\hat{v}^2$. As we do not believe in negative variances, we take this to mean that the slope has the smallest acceptable value, zero, which implies that the variance is constant, i.e. that $(\sigma_0^2/\sigma_2^2)_{ij}$ is very large. On the other hand, if the ordinate intercept is negative (Fig. 6.6(c)), the naively drawn straight line implies that the variance is negative at small $\hat{v}^2$. As we still do not believe in negative variances, we take this to mean that the intercept has the smallest acceptable value, zero, which implies that the variance is proportional to $\hat{v}^2$, i.e. that $(\sigma_0^2/\sigma_2^2)_{ij}$ is zero. This analysis explains the three cases expressed by eqn 6.7. Fortunately, we do not need to study the case where the slope and intercept are both negative, as this cannot arise.

It is not necessary to choose between the methods described in this and the preceding sections, because they can be used in conjunction, i.e. it is quite feasible both to adjust the weighting parameter σ_0^2/σ_2^2 or α and to use the biweight to adjust the weights after every iteration, following the general protocol set out in Fig. 6.7. If the weighting function is known independently the steps where it is estimated can be omitted, or if one feels that outlier protection is unnecessary the biweight adjustments can be omitted. When applied to Michaelis–Menten data (Cornish-Bowden and Endrenyi 1981), this approach works about as well as the median estimation described in Section 5.5, though it demands much more computation. Unlike median estimation, however, it is readily generalizable to more complex equations, such as inhibition (Cornish-Bowden and Endrenyi 1986a) and two-substrate equations (Cornish-Bowden and Endrenyi 1986b): any linear equation or any linearizable equation can be handled in essentially the same way.

Why does even this degree of restriction apply? Why cannot the method be applied to intrinsically non-linear equations? The problem is that in an inherently non-linear problem convergence is defined in terms of the minimization of a function, normally a sum of squares, and one knows that one is advancing towards a solution if the function value decreases from one iteration to the next. However, if the weighting function is variable, the sum of squares cannot be used as a termination criterion, because it has variable dimensions—the dimensions of v if σ_0^2/σ_2^2 is infinite (or if $\alpha = 0$), dimensionless if $\sigma_0^2/\sigma_2^2 = 0$ (or if $\alpha = 2$), and intermediate non-integral dimensions for intermediate weighting functions. There is no sense in which it can be said to decrease as the calculation proceeds and so some other criterion for termination is required. Provided

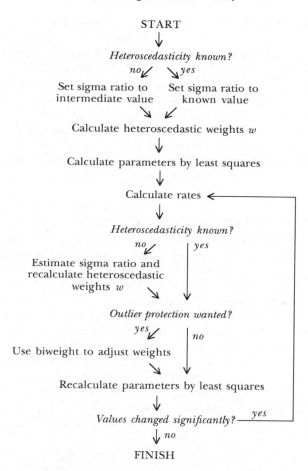

Fig. 6.7. *Flow chart for robust regression with no prior assumption about heteroscedasticity.*

that each iteration can be done exactly (for the particular weights defined after the previous iteration), self-consistency can be used as a termination criterion, but this effectively limits the approach to linear and linearizable models.

One complication may still arise, even with such models, because the weighting parameter may oscillate indefinitely between two values so that the solution is never self-consistent. This can be overcome by limiting the amount by which the weighting parameter is allowed to change in each iteration: for example, unlimited changes may be permitted in the first ten iterations, followed by tighter and tighter constraints in the subsequent ones.

A detailed example of the approach outlined in Fig. 6.7 is given in the next section. I consider that this is the method of choice for analysing enzyme kinetic data. Although more powerful methods may well exist, of course (see, for example, Ruppert *et al.* 1989; Nelder 1991), these require considerably deeper understanding of statistical theory than most biochemists aspire to, and so experimentalists who use them place themselves more in the hands of experts than is wise. The aim in this section has not been so much to describe a method that will give the best possible results in all circumstances, but to present one that will perform well nearly always, and is also easy to understand and easy enough to program for interested users to do it for themselves.

6.4   Worked example of robust regression

This section describes in detail how to use the procedure set out in Fig. 6.7 to fit the following set of ten (a, v) pairs to the Michaelis–Menten equation: $(1, 0.219)$, $(2, 0.343)$, $(3, 0.411)$, $(4, 0.470)$, $(5, 0.490)$, $(6, 0.525)$, $(7, 0.512)$, $(8, 0.535)$, $(9, 0.525)$, $(10, 0.540)$. As the objective is to describe where the numbers come from rather than why the calculation is done in that way, the presentation will be more terse than elsewhere in this book. To pad out the description with more explanation it will be necessary to refer to the description in the preceding two sections.

The hyperbolic equation is $v = Va/(K_m + a)$, which corresponds to the linear equation $y = b_1 x_1 + b_2 x_2$ with $y = 1/v$, $x_1 = 1/a$, $x_2 = 1$, $b_1 = K_m/V$, $b_2 = 1/V$. Thus the variables to be used as as given in Table 6.1.

The appropriate weights for y are obtained from those for v by multiplying by $\hat{v}^3 v$; however, as initially we have no values for $\hat{v}$ we use v^4 instead. If we assumed a constant variance for v the appropriate weight for v would be 1; if we assumed constant coefficient of variation it would

Table 6.1   Linear variables corresponding to a set of Michaelis–Menten data

a	v	x_1	x_2	y
1.000	0.219	1.00000	1.00000	4.56621
2.000	0.343	0.50000	1.00000	2.91545
3.000	0.411	0.33333	1.00000	2.43309
4.000	0.470	0.25000	1.00000	2.12766
5.000	0.490	0.20000	1.00000	2.04082
6.000	0.525	0.16667	1.00000	1.90476
7.000	0.512	0.14286	1.00000	1.95313
8.000	0.535	0.12500	1.00000	1.86916
9.000	0.525	0.11111	1.00000	1.90476
10.000	0.540	0.10000	1.00000	1.85185

be $1/\hat{v}^2$. However, as we have no reason to be sure of either of these we start with an intermediate weighting scheme in which heteroscedastic weights w_1 are calculated as $(\bar{v}^2 + \sigma_0^2/\sigma_2^2)/(\hat{v}^3/v + \sigma_0^2/\sigma_2^2)$. In the absence of a priori information about the value of σ_0^2/σ_2^2 it is set initially to $\bar{v}^2$, or 0.208 85. The numerator in this expression is constant and is just a scaling factor: it could be set to 1 without affecting the results, but the numerator given here ensures that values of w_1 span a convenient range on both sides of 1. In the denominator v is used as the initial estimate of $\bar{v}$ because the latter is initially unknown. The actual weights W used for the linear dependent variable y are calculated as $v^4 w_1 w_2$, in which v^4 compensates for the linear transformation, w_1 is the heteroscedastic weight, and w_2 is a biweight adjustment that is initially set to 1 for all observations. All of this leads to the values listed in Table 6.2.

The summations lead to the following pair of simultaneous equations for the parameter values: $2.0460 = 0.1966 b_1 + 0.9850 b_2$; $0.1966 = 0.0224 b_1 + 0.0883 b_2$. Their solutions $b_1 = 2.7465$ and $b_2 = 1.5273$ are now used to obtain calculated rates $\hat{v}$, which allow recalculation of w_1 with the formula given above and hence residuals $w_1^{1/2}|v - \hat{v}|$, as listed in Table 6.3.

The median absolute residual is

$$S = (0.008\,29 + 0.013\,34)/2 = 0.010\,81.$$

This is used to calculate the biweight factors $w_2 = (1 - u^2)^2$ shown in the last column of the table, u being a scaled residual calculated as $u = w_1^{1/2}|v - \hat{v}|/6S$. For example, in the top line, $u = 0.018\,73/0.6486 = 0.2888$, hence $w_2 = (1 - 0.2888^2)^2 = 0.8402$. (If any u values were greater than 1, w_2 would be set to zero, but there are no such cases in this example.)

Before these biweight factors are used, the weighting parameter σ_0^2/σ_2^2 will be estimated by considering the observations in pairs. The first pair of observations gives $v_1 = 0.219$, $\hat{v}_1 = 0.23398$, $v_1 - \hat{v}_1 = -0.01498$, $v_2 = 0.343$, $\hat{v}_2 = 0.34476$, $v_2 - \hat{v}_2 = -0.00176$, from which we can estimate a slope σ_2^2 of $(0.014\,98^2 - 0.001\,76^2)/(0.233\,98^2 - 0.344\,76^2) = -0.003\,45$. As this slope is negative we do not calculate an ordinate but treat it as giving an infinite estimate of σ_0^2/σ_2^2. The first pair of observations to give positive values for both slope and intercept is the first and fourth, with $v_1 = 0.219$, $\hat{v}_1 = 0.233\,98$, $v_1 - \hat{v}_1 = -0.014\,98$, $v_4 = 0.470$, $\hat{v}_4 = 0.45169$, $v_4 - \hat{v}_4 = 0.018\,31$; these give a slope $\sigma_2^2 = (0.014\,98^2 - 0.018\,31^2)/0.233\,98^2 - 0.451\,69^2) = 0.000\,742\,6$ and intercept $\sigma_0^2 = 0.014\,98^2 - 0.000\,742\,6 \times 0.233\,98^2 = 0.000\,183\,8$, and hence $\sigma_0^2/\sigma_2^2 = 0.2475$. Proceeding in this way through all pairs of observations yields the 45 estimates of σ_0^2/σ_2^2 listed in Table 6.4.

Each of the entries shown as $> \infty$ corresponds to a negative calculated

Table 6.2 Calculation of preliminary estimates

w_1	w_2	W	Wy^2	Wx_1y	Wx_2y	Wx_1^2	Wx_1x_2	Wx_2^2
1.6265	1.0000	3.7×10^{-3}	0.0780	0.0171	0.0171	3.7×10^{-3}	3.7×10^{-3}	3.7×10^{-3}
1.2793	1.0000	0.0177	0.1505	0.0258	0.0516	4.4×10^{-3}	8.9×10^{-3}	0.0177
1.1057	1.0000	0.0316	0.1868	0.0256	0.0768	3.5×10^{-3}	0.0105	0.0316
0.9720	1.0000	0.0474	0.2147	0.0252	0.1009	3.0×10^{-3}	0.0119	0.0474
0.9304	1.0000	0.0536	0.2234	0.0219	0.1095	2.1×10^{-3}	0.0107	0.0536
0.8622	1.0000	0.0655	0.2376	0.0208	0.1258	1.8×10^{-3}	0.0109	0.0655
0.8868	1.0000	0.0609	0.2325	0.0170	0.1190	1.2×10^{-3}	8.7×10^{-3}	0.0609
0.8437	1.0000	0.0691	0.2415	0.0161	0.1292	1.1×10^{-3}	8.6×10^{-3}	0.0691
0.8622	1.0000	0.0655	0.2376	0.0139	0.1248	8.1×10^{-4}	7.3×10^{-3}	0.0655
0.8346	1.0000	0.0710	0.2434	0.0131	0.1314	7.1×10^{-4}	7.1×10^{-3}	0.0710
Sum			2.0460	0.1966	0.9850	0.0224	0.0883	0.4861

Table 6.3   Calculation of biweight factors

v	$\hat{v}$	w_1	$w_1^{1/2}\lvert v - \hat{v}\rvert$	w_2
0.21900	0.23398	1.56241	0.01873	0.8402
0.34300	0.34476	1.27223	0.00199	0.9981
0.41100	0.40937	1.11160	0.00172	0.9986
0.47000	0.45169	1.03155	0.01860	0.8423
0.49000	0.48156	0.95638	0.00826	0.9679
0.52500	0.50377	0.92336	0.02040	0.8119
0.51200	0.52093	0.86133	0.00829	0.9676
0.53500	0.53459	0.84484	0.00038	0.9999
0.52500	0.54571	0.80574	0.01859	0.8424
0.54000	0.55496	0.79508	0.01334	0.9172

slope, and each of those shown as < 0 corresponds to a negative calculated intercept. As there are (by chance) exactly 20 of each, the median is the median of the remaining five values, i.e. 0.1426. For as simple an example as this it is quite feasible to calculate the median exactly, but in examples with large numbers of observations this implies a very large amount of calculation and, more important, the need to store a large number of values while the median is being found. All this would serve no great purpose, as an approximate calculation is quite adequate. This can be done by counting the number of values of σ_0^2/σ_2^2 that fall within predefined ranges. A difficulty, however, is that it is not very convenient to interpolate on a scale in which the values range from 0 to infinity, with substantial numbers of results at the extremes. It is convenient, therefore, to 'stabilize' the σ_0^2/σ_2^2 values by converting them to corresponding α values as $2\bar{v}^2/(\bar{v}^2 + \sigma_0^2/\sigma_2^2)$, estimating the median α, and then converting back to σ_0^2/σ_2^2. Each of the values of $\sigma_0^2/\sigma_2^2 > \infty$ corresponds to $\alpha < 0$; each of the values of $\sigma_0^2/\sigma_2^2 < 0$ corresponds to $\alpha > 2$; the remainder give $\alpha = 0.9153$ for $\sigma_0^2/\sigma_2^2 = 0.2475$, $\alpha = 1.1883$ for $\sigma_0^2/\sigma_2^2 = 0.1426$, $\alpha = 0.9926$ for $\sigma_0^2/\sigma_2^2 = 0.2120$, $\alpha = 1.2269$ for $\sigma_0^2/\sigma_2^2 =$

Table 6.4   Estimates of the sigma ratio

2	3	4	5	6	7	8	9	10	
$> \infty$	$> \infty$	0.2475	$> \infty$	0.1426	$> \infty$	$> \infty$	0.2120	$> \infty$	1
	$> \infty$	< 0	< 0	< 0	< 0	$> \infty$	< 0	< 0	2
		< 0	< 0	< 0	< 0	$> \infty$	< 0	< 0	3
			$> \infty$	< 0	$> \infty$	$> \infty$	0.1316	$> \infty$	4
				< 0	0.1021	$> \infty$	< 0	< 0	5
					$> \infty$	$> \infty$	$> \infty$	$> \infty$	6
						$> \infty$	< 0	< 0	7
							< 0	< 0	8
								$> \infty$	9

Table 6.5 Recalculation of sums

w_1	w_2	W	Wy^2	Wx_1y	Wx_2y	Wx_1^2	Wx_1x_2	Wx_2^2
1.6392	0.8402	3.9×10^{-3}	0.0806	0.0176	0.0176	3.9×10^{-3}	3.9×10^{-3}	3.9×10^{-3}
1.3018	0.9981	0.0183	0.1552	0.0266	0.0532	4.6×10^{-3}	9.1×10^{-3}	0.0183
1.1220	0.9986	0.0316	0.1870	0.0256	0.0769	3.5×10^{-3}	0.0105	0.0316
1.0343	0.8423	0.0377	0.1708	0.0201	0.0803	2.4×10^{-3}	9.4×10^{-3}	0.0377
0.9529	0.9679	0.0505	0.2102	0.0206	0.1030	2.0×10^{-3}	0.0101	0.0505
0.9175	0.8119	0.0500	0.1814	0.0159	0.0952	1.4×10^{-3}	8.3×10^{-3}	0.0500
0.8515	0.9676	0.0596	0.2275	0.0166	0.1165	1.2×10^{-3}	8.5×10^{-3}	0.0596
0.8341	0.9999	0.0682	0.2382	0.0159	0.1274	1.1×10^{-3}	8.5×10^{-3}	0.0682
0.7929	0.8424	0.0570	0.2068	0.0121	0.1086	7.0×10^{-4}	6.3×10^{-3}	0.0570
0.7817	0.9172	0.0662	0.2269	0.0123	0.1226	6.6×10^{-4}	6.6×10^{-3}	0.0662
Sums			1.8846	0.1833	0.9013	0.0214	0.0814	0.4429

Table 6.6 Progress of the calculation

Iteration	b_1	b_2	σ_0^2/σ_2^2	$\sum w_2$
1	2.7465	1.5273	0.20885	10.00
2	2.7675	1.5265	0.17674	9.19
3	2.8029	1.5198	0.08602	9.21
4	2.8221	1.5159	0.05494	9.22
5	2.8221	1.5158	0.05494	9.23
6	2.8220	1.5158	0.05494	9.23

0.1316, and $\alpha = 1.3432$ for $\sigma_0^2/\sigma_2^2 = 0.1021$. Classification into six bins shows that 20 pairs of observations give α less than 0, 0 give values in the range 0–0.5, 2 give values in the range 0.5–1, 3 give values in the range 1–1.5, 0 give values in the range 1.5–2, and 20 give values greater than 2. The median must therefore be the first of the three values in the range 1–1.5. If we did not have the actual values, we might hypothesize that each lay in the middle of a range obtained by dividing the range 1–1.5 into three subranges, i.e. we could estimate the median as $1 + 0.5/6$, or 1.0833. This can be converted back into a value of $\sigma_0^2/\sigma_2^2 = 0.1767$, which is the value of σ_0^2/σ_2^2 to be used for calculating weights in the next iteration, to produce the summation given in Table 6.5.

The summations lead to the following pair of simultaneous equations for the parameter values: $1.8846 = 0.1833b_1 + 0.9013b_2$, $0.1833 = 0.0214b_1 + 0.0814b_2$, with solutions $b_1 = 2.7675$, $b_2 = 1.5265$.

From this point the calculation is just a repetition of the steps already taken until the results are self-consistent, with successive estimates of the parameters and weighting functions as shown in Table 6.6.

6.5 The jackknife and bootstrap

The jackknife is one of a number of statistical techniques that are useful for problems that have no known analytical solutions, or where the analytical solutions are too difficult to apply to be practical. This may happen, for example, in cases where the parameters of interest enter into the fitted equation in a very complicated way (e.g. Cornish-Bowden and Wong 1978). It was first introduced as a way of eliminating a particular kind of bias, but is now much more widely used for obtaining reasonably believable statistical information when one cannot trust the more usual methods. The basic idea may be illustrated by considering the mean $\bar{x}$ of a set of n numbers, which satisfies the equation $n\bar{x} = \sum x$. Suppose we define $\bar{x}_{-j}$ as the mean of the same sample omitting the jth observation, then this must, of course, satisfy the equation $(n-1)\bar{x}_{-j} = -x_j + \sum x$, and so, in the unlikely event that we happened to know $\bar{x}$ and $\bar{x}_{-j}$ but had lost the value of x_j, we could calculate it from the following relationship:

$$x_j = n\bar{x} - (n-1)\bar{x}_{-j}. \qquad (6.8)$$

This may seem rather pointless, as it is hardly likely that we would know $\bar{x}$ and $\bar{x}_{-j}$ but not x_j. The beauty of it, however, is that it can be applied to any estimator at all, not just to a sample mean. In other words, if $\hat{b}_0$ is an estimate of some parameter in a multi-parameter model obtained using all the observations, and $\hat{b}_{-j}$ is the estimate obtained in the same way but using all except the jth observation, then we can define a *pseudo-value* $\tilde{b}_j$ as follows:

$$\tilde{b}_j = n\hat{b}_0 - (n-1)\hat{b}_{-j}. \qquad (6.9)$$

We call this a 'pseudo-value', and give it a special symbol, because it is not in fact an observed value of anything. Nonetheless, the analogy with eqn 6.8 suggests that we could treat it *as if it were* a direct observation of b. As we can in general calculate n pseudo-values, we have in effect a way of artificially generating a sample of n pseudo-observations. We can, for example, calculate a mean of the pseudo-values,

$$\tilde{b}_0 = \frac{1}{n} \sum \tilde{b}_i, \qquad (6.10)$$

and calculate a standard error of the mean in the ordinary way. This process of converting $\hat{b}_0$ into $\tilde{b}_0$ is called *jackknifing*.

Although the bias-removing effect of the jackknife is not now regarded as its principal claim to attention, it is perhaps worth examining, as the theory is quite straightforward and it may still have some application in some circumstances. Suppose that an estimator $\hat{b}_0$ of some parameter β is known to have a bias that tends to zero as the sample size increases, so that its expected value $E(\hat{b}_0)$ is not β but $\beta + A/n$, where A is a constant, then $E(\hat{b}_{-i})$ is clearly $\beta + A/(n-1)$, and so

$$E(\tilde{b}_i) = nE[\hat{b}_0] - (n-1)E[\hat{b}_{-i}] = n\beta + A - (n-1)\beta - A = \beta. \quad (6.11)$$

An example of an estimator that is biassed in exactly this type of way is the sum of squares of deviations in linear least squares when it is regarded as an estimator of the sum of squares of errors (cf. eqn 1.60). In this case the magnitude of the bias is known and can easily be corrected by a simple multiplication, but for other parameters the magnitude of the correction may be unknown. As division by an estimate (or any value that is not known exactly) is a simple way to introduce bias, and as many of the parameters of interest to biochemists, such as K_m, are obtained by division, bias can be expected to be a problem in kinetics whenever the errors are large.

Several warnings about jackknifing should be made, of which the first ought to be obvious. As jackknifing any parameter estimate involves repeating the entire initial fitting procedure for each observation, it requires about $(n + 1)$ times as much computation as making the initial estimate, not counting the computation involved in processing the pseudo-values at the end. It may thus involve a considerable amount of computer time, and so any program capable of performing a jackknife needs to include a straightforward way of abandoning the computation in the middle. Second, although most kinds of estimator can usefully be jackknifed, there are some exceptions, of which the simplest and most obvious is the ordinary median, as one may see by noting that the pseudo-values obtained by jackknifing the median of the values $(1, 2, 3, 4, 5, 6, 7, 8, 9, 10)$ would be $(1, 1, 1, 1, 1, 10, 10, 10, 10, 10)$. A third point is that when good methods already exist for estimating a parameter of interest, jackknifing it may simply replace a good estimate by another with an appreciably larger variance.

The jackknife is the simplest of a class of statistical techniques known as *resampling*, which set out to obtain from sets of data additional samples that might have been observed. The *bootstrap* is a more complex and more powerful example of this idea. As it has been little used in enzyme kinetics to date it will be described very briefly, but it is potentially a valuable technique. The essential idea is as follows. After a set of data has been fitted to an equation, the hypothesis may be proposed that the model fitted is the true model and the fitted parameters are the true parameters, so the calculated rates are the true rates and the calculated deviations are the true errors. After allowing for any heteroscedasticity built into the fitting procedure it is reasonable to say that one error is just as good as another, and that it would have been just as likely for the same errors to have been differently distributed among the same true rates. In other words a new set of data can be created by permuting the deviations, with, according to our hypothesis, exactly the same statistical properties as the original set of data. For n observations there will be $n!$ different ways of doing this, which will usually be many more than enough to generate a large sample of experiments, so that it will usually be sufficient to generate, say, 1000 pseudo-experiments with random permutations of the original errors. Each of these pseudo-experiments can then be analysed as if it were a real experiment, thereby generating information about the reproducibility of the parameter estimates.

6.6   Minimax fitting

Strictly speaking, minimax fitting does not belong in a book about data
analysis at all, because it is a technique from numerical analysis rather
than statistics. However, both disciplines are concerned with equation
fitting, and some of the same computational techniques are applicable
(virtually all search techniques, for example), and occasionally biochemists
have equation-fitting problems that demand (though they usually do not
get) solutions from numerical analysis. Perhaps most important, biochemists occasionally consult numerical analysts under the impression that
they are consulting statisticians, and if neither party to the discussion
realizes that there is a misunderstanding the advice given may be entirely
inappropriate.*

The essential difference between the two is that although both disciplines are concerned with minimizing the effects of error, statistical analysis assumes that the error is random, whereas numerical analysis assumes
that it arises from fitting an incorrect equation. Why would one ever
deliberately choose to fit an equation knowing that it was incorrect? One
reason is that sometimes the true function is difficult and complicated to
compute, whereas a function that is quick and easy to compute generates
almost exactly the same values. This is the case, for example, for the
dependence of a metabolic flux on the catalytic activity of a single enzyme:
computing such a flux exactly requires a specialized computer program,
but it can usually be approximated quite well by the Michaelis–Menten
equation. Modelling of the cooperative kinetic behaviour of a multi-
subunit enzyme theoretically requires a multi-parameter equation, but it
can often be modelled surprisingly accurately with the Hill equation. It
may also happen that the true equation is simply unknown, even though
abundant data exist for knowing the behaviour accurately over a wide
range of conditions.

In all of these cases it may be useful to fit an equation that is not derived
from any theoretical model but which happens to fit sufficiently well. The
question then arises as to how the goodness of fit ought to be judged. For
the reasons considered already in this book, statistical methods generally
try to minimize some *average* of the estimation error, but this is by no
means appropriate if the error is not statistical in origin and we want to
use the fitted equation for predicting the behaviour at some point where
measurements are not available. What use is it to know that the average

---

*The term *orthogonal polynomials* is a useful diagnostic. If your expert mentions these
you are probably talking to a numerical analyst, not a statistician. Orthogonal polynomials
have some value for deriving equations that are to be used for prediction, but almost none for
problems of model discrimination, which is what most biochemists are interested in most of
the time.

error in some output variable y is less than 1 per cent over a range from 0
to 100 of some input variable x if we only ever need values of x in the
range 5 to 10? However good the average may be, it is not much use if the
error in the region of interest is much higher. So clearly we need to specify
a range of interest, with the implication that we do not care how bad the
fit may be elsewhere. Even within this range an average is far from ideal,
because the point of worst fit may happen to coincide with the point of
greatest interest. What is needed is a way to optimize the worst case. We
would normally prefer a fitted function $f(x)$ that agrees with a true
function $\phi(x)$ to better than 1 per cent over the whole of the range of x
of interest over a function that agrees on average to better than 0.1 per
cent but deviates sharply at one end of the range with errors exceeding 25
per cent.

It follows, then, that the appropriate criterion of fit in such cases is a
minimax criterion, i.e. minimization of the maximum absolute deviation.
As this criterion shares with least-absolutes fitting the use of absolute
values, it tends to be non-smooth and responds badly to fitting methods
that depend on smooth behaviour of derivatives, though search methods
normally work well.

It remains to consider what sort of easily computed equations are likely
to give a good fit to the kinds of curves that occur in biochemistry. For
simple curves, i.e. curves that are almost straight lines but show some
curvature, a quadratic equation will often work well, and is easy to fit by
least squares because it is linear (although, for the reasons given above,
one ought not to be satisfied with a fit that minimizes an average
deviation). However, one should resist the temptation to generalize further
and just add terms in x^3, x^4, etc. to improve the fit.

There are several reasons why it is not a good idea to fit simple
polynomials of increasing order. The first—thoroughly discussed in textbooks of numerical analysis—is that each time a term is added to a simple
polynomial *all* of the coefficients already calculated change, normally
quite drastically, so parameters cannot just be added to a known model
calculating one new parameter at each step. More seriously, simple polynomials suffer very severely from problems of arithmetic, because as soon
as there are more than three or four terms the whole calculation comes to
be dominated by steps involving differences between large numbers. Both
of these difficulties can be overcome by making the polynomials *orthogonal*, but as there are additional objections that make any kind of
polynomial fitting (beyond the quadratic or cubic stages) inappropriate for
biochemical use, I shall not discuss this refinement here.

A different kind of danger with polynomials (whether orthogonal or
not), especially if they are applied to experimental data containing significant statistical error, is that the more terms there are the more one tends to
find significance in insignificant fluctuations, and it becomes difficult or

impossible to distinguish between fluctuations that are artefacts of measurement error and real variations in the true function. Again, however, one is usually safe from this kind of problem if one does not go beyond cubic equations.

Finally, the curves that occur in enzyme kinetics often take the form of transitions between simple asymptotes. For example, the Michaelis–Menten equation can be thought of as a transition from a proportional dependence of rate on concentration at low concentrations to independence of rate from concentration at high concentrations. Polynomials are in general very bad at representing the approach to an asymptote. For example, Philo and Selwyn (1973) found that even an 11th-order polynomial could not give an acceptable fit to a progress curve calculated exactly from the integrated Michaelis–Menten equation. One can improve the behaviour in this regard very substantially by including terms in negative powers in the polynomial, or by using a rational function (i.e. the ratio of two polynomials) instead of a polynomial. Polynomials with negative powers are linear equations, and are thus in principle easy to fit, but they are less versatile for dealing with a wide range of curves than rational functions, which are not linear.

6.7   Reading the statistics literature

Although I have tried in this book to cover the statistical theory needed for understanding the methods that are in use for fitting the common enzyme kinetic equations to data, there will inevitably be some omissions that will lead some readers to want to consult the statistics literature, both textbooks and periodicals. Unfortunately it is quite difficult to find textbooks at an appropriate level. There is an abundant over-supply of elementary texts, but these usually confine what little they have to say about model fitting to a chapter on fitting the straight line by least squares, with no mention of the assumptions implicit in this approach, and no mention of the fact that many of the equations met with in real life (including virtually all of those of interest to the biochemist) are non-linear.

Excellent advanced textbooks, such as that of Kendall and Stuart (1969), are of course available as well, but these demand a much higher grasp of mathematics than is needed for the present book, and readers able to use them will hardly need this book as an introduction. An authoritative account of classical regression (mainly linear least squares) is provided by Draper and Smith (1981), but a more iconoclastic view of regression closer to the spirit of the present book is that of Mosteller and Tukey (1977), which should be read with the companion book of Tukey (1977). As a rough but reliable guide it is fair to say that *anything* written by Tukey will contain gems of wisdom expressed in a language that is by no means too technical for the non-statistician.

In the third category are statistics books written in relation to specific applications. There are a great many of these, but very few are particularly relevant to biochemistry (which in this respect must be regarded as an entirely different field from biology, especially classical genetics). The book edited by Endrenyi (1981) contains several chapters of great interest, of which that of Watts (1981) may be warmly recommended as presenting the view of a statistician with a good knowledge of the sort of problems that interest biochemists and pharmacologists. Although now quite old, the book by Colquhoun (1973) still provides probably the best general introduction to biological statistics.

So far as periodicals are concerned, articles on problems of importance to biochemists are not at all abundant, but they appear occasionally in such journals as *Biometrics*, *Technometrics*, and the *Journal of the Royal Statistical Society*, *Series C* (which is also called *Applied Statistics*). When reading these, one should not forget that although one can usually take the authors' grasp of statistics as beyond question, their knowledge of biochemistry may be much more dubious. This is fairly easy for the biochemist reader to recognize, but a more subtle difficulty arises when unfamiliar symbols and terminology are used to present familiar ideas.

An example is provided by a paper of Raaijmakers (1987), which presents the model to be fitted as follows:

$$v_i = \frac{\sigma s_i}{s_i + \beta} + \frac{s_i}{s_i + \beta} \varepsilon_i. \tag{6.12}$$

It is not difficult to recognize this as the Michaelis–Menten equation with (implicitly in the equation but explicit in the text) a constant coefficient of variation. It is much less obvious, however, that the proposed estimate of the Michaelis constant,

$$\hat{\beta} = \frac{\overline{X} \sum (Y - \overline{Y})^2 - \overline{Y} \sum (X - \overline{X})(Y - \overline{Y})}{\overline{Y} \sum (X - \overline{X})^2 - \overline{X} \sum (X - \overline{X})(Y - \overline{Y})}, \tag{6.13}$$

is identical with what one would get by dividing the expression in eqn 2.13 by that in eqn 2.12,

$$\hat{K}_m = \frac{\sum v^2 (\sum v/a) - (\sum v^2/a) \sum v}{(\sum v^2/a^2) \sum v - (\sum v^2/a)(\sum v/a)}, \tag{6.14}$$

if one allows for the substitutions $X_i = v_i/s_i$, $Y_i = v_i$.

It is not surprising that they should be identical, of course, because they have been derived from the same initial assumption, and although

Raaijmakers describes his method as based on maximum likelihood, it assumes Gaussian errors and thus gives the same estimate as least squares. The derivation of eqns 2.12 and 2.13 follows that of Johansen and Lumry (1961), and Raaijmakers was mistaken in thinking that his result was novel. Moreover, his claim that it is superior to published methods was also mistaken, as it depended on the truth of the starting assumption. The whole basis of median estimation (and of robust estimation in general) is that the proper error assumptions are *not known*, and that it is wiser to choose a method that will always perform reasonably well, even though it may be sub-optimal in the imaginary world where the true error distribution is known. No sensible proponent of robust methods would claim that they perform better than classical methods when the error distribution is known, and consequently no sensible proponent of robust methods will be surprised to learn of a specific set of assumptions that allows least squares to perform better.

The paper of Raaijmakers (1987) illustrates another question that the biochemist needs to ask when reading the statistics literature: how well do the assumptions on which the analysis is based conform to reality in biochemistry? Raaijmakers considers two levels of error: one, described as 'small', had a coefficient of variation of 0.05, which is reasonable enough, though rather high when one remembers that it implies about 5 per cent of observations with deviations of 10 per cent or more; the other, described as 'relatively large', had a coefficient of variation of 0.25, which can only be regarded as a case of testing a method to destruction, as it certainly has little relevance to modern enzyme kinetics. The sample sizes were 5, 10, 20, 50, 100, 200, and 500, but only five substrate concentrations were considered, namely $0.1\,K_m$, $0.4\,K_m$, $1.6\,K_m$, $6.4\,K_m$, and $25.6\,K_m$, so that, for example, the experiment with 500 observations consisted of 100 replications at each of these concentrations. Again, any similarity between these simulations and the sort of enzyme kinetic measurements actually carried out by biochemists can only have been by chance.

II Interlude

7
Analysis of an example

7.1 Introduction: acylaminoacyl-peptidase

Some experiments on acylaminoacyl-peptidase provide a practical example of data analysis in enzyme kinetics. They illustrate a typical problem of data analysis, where one has first to assess how much information, both quantitative and mechanistic, can be deduced from experiments that are already finished by the time one sees them, and second to suggest how future experiments might be planned so as to provide more complete information.

Acylaminoacyl-peptidase (EC 3.4.19.1) is found in mammalian tissues such as the liver and intestinal mucosa, and catalyses the hydrolysis of N-acetyl or N-formyl peptides, with release of the N-terminal N-acylamino-acid. The data of Raphel (1994) on the enzyme from pig intestinal mucosa consist of measurements of the inhibition of the hydrolysis of acetyl-L-alanine p-nitrophenyl ester by acetyl-D-alanine and by acetyl-L-alanine. However, although the specific example refers to enzyme inhibition, the principles that ought to emerge from study of the experiments are applicable to many other kinds of experiment in enzyme kinetics as well.

This chapter provides a link between the theoretical development of the preceding chapters and the user's manual for the computer program Leonora of the later ones. It can be regarded either as a postscript to Part I or as a prelude to Part III. All of the statistical results to be given were obtained with the program Leonora, but in principle similar results could be obtained with other suitable programs. To avoid making this book impossibly long, reference will be made to many more plots and tables of results than can conveniently be shown here: all of these can easily be generated on the computer screen with a program such as Leonora.

To some extent the analysis to be given is rather similar to parts of Chapter 3, but the emphasis is quite different: there we used an artificially constructed example to illustrate specific points made in the course of a primarily theoretical discussion; here the objective is to apply the ideas discussed in the earlier chapters to a real example, taking the data exactly as they were supplied by the experimenter.

Table 7.1   Inhibition of acylaminoacyl-peptidase by acetyl-L-alanine

The inhibition was studied at pH 8 with acetyl-L-alanine p-nitrophenol as substrate. The substrate and inhibitor concentrations are represented by a and i respectively. In the discussion in the text the rates are taken as being numerically 10^4-fold smaller than those listed (for reasons discussed in Section 7.2).

a (mM)	i (mM)	v (nmol min^{-1})	a (mM)	i (mM)	v (nmol min^{-1})
0.5	0.0	17273	2.0	0.01	46364
0.5	0.05	16364	2.0	0.03	40909
0.5	0.1	13636	2.0	0.06	40000
0.5	0.5	8182	2.0	0.1	39091
0.5	0.3	10000	2.0	0.3	32727
0.5	0.8	5000	2.0	0.5	28182
0.5	1.0	4878	2.0	1.0	20000
1.0	0.0	32273	2.0	1.5	13000
1.0	0.1	27273	2.0	2.0	11818
1.0	0.3	20000	2.0	2.5	8482
1.0	0.5	15300	2.0	3.0	7727
1.0	0.7	10909	2.0	3.5	6364
1.0	1.0	10000	2.0	4.0	5494
1.0	1.2	9091	2.0	4.5	4950
1.0	1.5	8182	2.0	5.0	4090
2.0	0.0	46364	2.0	6.0	3636

7.2 Preliminary examination of the data

Two sets of data for hydrolysis of acetyl-L-alanine p-nitrophenyl ester catalysed by acylaminoacyl-peptidase are listed in Tables 7.1 (inhibition by acetyl-L-alanine) and 7.2 (inhibition by acetyl-D-alanine); there are 32 rate measurements in the former case and 27 in the latter.

There are two aspects of the data that should attract the attention before one even begins to analyse them, as they could be symptomatic of mistakes. The first is that although in general the observations are in order of increasing inhibitor concentration for each group at the same substrate concentration, there is one pair of observations (the fourth and fifth in the left-hand column of Table 7.1) where this sequence is reversed. As it happens there is no mistake here, but one should always check whenever any suspicious characteristics are noticed.

A second surprising point is that some of the observations in Table 7.1 are expressed with one significant figure, whereas others are expressed to five, with very few intermediate cases; in Table 7.2 most of the values appear to have seven or eight significant figures, though a few have only one. Moreover, there are two identical entries of 46 364 nmol min^{-1} in Table 7.1, and 12 269.939 and 23 640.662 nmol min^{-1} occur twice each in Table 7.2, coincidences that one would expect to observe very rarely

Table 7.2 Inhibition of acylaminoacyl-peptidase by acetyl-D-alanine
Details are as for Table 7.1.

a (mM)	i (mM)	v (nmol min^{-1})	a (mM)	i (mM)	v (nmol min^{-1})
0.5	0.0	16949.153	1.0	12.0	12722.646
0.5	2.0	13642.565	1.0	15.0	10000.000
0.5	4.0	12269.939	1.0	20.0	7727.975
0.5	6.0	10905.125	2.0	0.0	66666.667
0.5	8.0	8271.299	2.0	4.0	33222.591
0.5	10.0	7955.449	2.0	5.0	29069.767
0.5	12.0	7501.875	2.0	6.0	23640.662
0.5	15.0	6816.633	2.0	8.0	23640.662
0.5	20.0	5227.392	2.0	10.0	20000.000
1.0	0.0	27247.956	2.0	12.0	17241.379
1.0	4.0	18181.818	2.0	15.0	13642.565
1.0	6.0	19083.969	2.0	17.0	12269.939
1.0	8.0	17035.775	2.0	20.0	11001.100
1.0	10.0	14992.504			

with numbers measured to such precision even if the experimental conditions are the same or almost the same. The large numbers of times in which the same sequences of digits occur (909, for example), also suggest a question that one should try to answer before proceeding further.

Further examination reveals that most or all of the rates in Table 7.1 can be expressed as simple fractions rounded to four decimal places. Thus, the numerical values of the first five rates are 190 000/11, 180 000/11, 150 000/11, 90 000/11, and 110 000/11 respectively. The prevalence of the number 11 in these cases suggests that the measurements were originally expressed in other units but have been scaled by dividing by 11 or a simple multiple of 11, an impression confirmed by the fact that most of the rates in the table yield integer multiples of 1000 if multiplied by 2.2.

This interpretation suggests that the original values (before scaling) were expressed to rather low precision (90 000 ± 10 000 nmol min^{-1} implies a precision of worse than 11 per cent, for example), and as rates can often be measured more accurately than this, it would be worthwhile enquiring whether more precise values are available. In the present discussion, however, we shall stay with the values as given, to avoid the need for further tables of data.

The problem here is not the scaling as such, but the implication that the original numbers were not as precise as they could have been. By contrast, scaling to different units is a perfectly reasonable procedure that should not lead to significant loss of precision, and should not affect the qualitative interpretation of the experiment. Nonetheless, the values in both tables are numerically very large, and this is undesirable for visual inspection of the data, as it can make it more difficult to recognize errors (e.g. if one writes $4.27 \times 3.92 = 1.674$ it is obvious from first inspection that this is wrong by a factor of ten, but the same error is by no means equally obvious if the equation is written $42\,700 \times 39\,200 = 16\,738\,000$). Moreover, numbers that differ greatly from 1 can lead to numerical inaccuracies even in the computer, and although this ought not to be a problem with a properly written modern program, one cannot always be certain that the computer programs one uses have been written with careful attention to such numerical considerations. For both of these reasons it is usually a good idea to scale the measurements so that non-zero values fall as far as possible in the range 0.1 to 10. In the remainder of this chapter I shall discuss the data in Tables 7.1 and 7.2 as if all the rates had been expressed in μmol min^{-1}, i.e. as if they were numerically 10^3-fold smaller than the values actually listed.

From the purely numerical point of view, a scaling factor of 10^4 would be better than 10^3, but 10^3 is an adequate compromise between the demands of numerical analysis and the need to express the results in biochemically meaningful terms. In any case, it is important to express all final results in meaningful units even if units chosen from the point of view of numerical analysis are used during the actual calculation, so that one can judge whether they are reasonable or not. For example, if an inhibitor produces an easily visible effect at a concentration of 5 mM, a calculation that yields a single inhibition constant of 1000 mM is obviously in error, but if all the values have been scaled to arbitrary units and then left in these units at the end, such an error could easily pass unnoticed.

Although the patterns of numbers in the two tables are superficially similar, the explanation for Table 7.2 appears to be different from that for Table 7.1. Instead of uniform scaling, these numbers appear to be reciprocals of numbers with two or three significant figures, i.e. $1/0.059 = 16.949\,153$, $1/0.0733 = 13.642\,565$, etc. This suggests that the rates originally measured have been converted to reciprocals and rounded to three decimal places for convenient plotting in double-reciprocal plots and then converted back into rates by taking reciprocals again. This is not a desirable process as it can result in loss of information: for example, 16.82 differs from 16.949\,153 by about 0.75 per cent, but its reciprocal is still 0.059 if written with two significant figures. Although such a loss of precision may be trivial for most purposes, it is an *avoidable* loss of precision that serves no purpose, and it ought therefore to be avoided.

A final point about the preliminary examination of the data is that once an explanation for any pattern in the numbers is found, one can check whether it accounts for *all* of the values: for example (Table 7.1), 48.78 yields 107.316 when multiplied by 2.2, which is not approximately an integer. Any such discrepancies should be checked with the original data, as they may be symptoms of errors of transcription.

As a number of distinct points have been introduced in this section it may be helpful to repeat them here: (i) the table of data should be checked for any oddities of sequence, large variations in apparent precision, or unexpected patterns in the numbers; (ii) if any explanations of these oddities suggest themselves one should check whether they imply an avoidable loss of precision; (iii) one cannot assume that an explanation of patterns in one set of data necessarily applies to superficially similar patterns in another, even if both sets of data come from the same source; (iv) if necessary the values should be scaled before analysis to bring them into a range that minimizes the likelihood of numerical inaccuracies; (v) if any patterns apply to most but not all of the values one should check whether the exceptions are the result of mistakes. The very length of this list is interesting: it emphasizes that there is quite a lot of useful work to do even before the computer is switched on!

7.3 Inhibition by acetyl-L-alanine

We now consider how well the data of Tables 7.1 and 7.2 agree with the ordinary models of enzyme inhibition, as exemplified by the equation for mixed inhibition:

$$v = \frac{Va}{K_m(1 + i/K_{ic}) + a(1 + i/K_{iu})} + e. \qquad (7.1)$$

This is the same as eqn 3.20, and it represents the rate v at concentrations a of substrate (acetyl-L-alanine p-nitrophenyl ester) and i of inhibitor (acetyl-L-alanine or acetyl-D-alanine) in terms of two Michaelis–Menten parameters V and K_m and two inhibition constants K_{ic} and K_{iu}. If K_{iu} is very large the term i/K_{iu} is negligible and the inhibition is competitive:

$$v = \frac{Va}{K_m(1 + i/K_{ic}) + a} + e. \qquad (7.2)$$

If K_{ic} is very large the term i/K_{ic} is negligible and the inhibition is uncompetitive:

$$v = \frac{Va}{K_m + a(1 + i/K_{iu})} + e. \qquad (7.3)$$

Fitting the data of Table 7.1 to the eqn 7.2 by the method schematized in Fig. 5.7 yields the following parameter values: $V = 98.4 \pm 9.2$ μmol min^{-1}, $K_m = 2.19 \pm 0.27$ mM, $K_{ic} = 0.315 \pm 0.023$ mM. Examination of the table of observed and calculated rates reveals no obvious anomalies, such as grossly large errors that might point to typing errors during entry

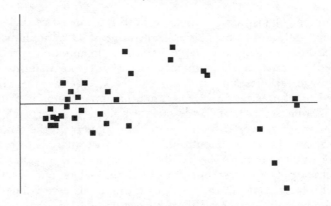

Fig. 7.1. *Residual plot for inhibition by acetyl-L-alanine.* The axes are left unlabelled to avoid distracting the eye with extraneous information (see Section 3.5). The ordinate axis shows the weighted difference between v and $\hat{v}$ in arbitrary units; the abscissa axis shows $\hat{v}$ on a scale that extends from zero to 50 μmol min^{-1}.

of the data. The plot of residual errors against calculated rate (Fig. 7.1), shows a predominantly random scatter of points, with enough of a suggestion of an inverted U to make it sensible to look at other models.

Fitting the same data to the equation for mixed inhibition, eqn 7.1, yields $V = 97.8 \pm 10.7$ μmol min^{-1}, $K_m = 2.18 \pm 0.38$ mM, $K_{ic} = 0.324 \pm 0.049$ mM, $K_{iu} = 15 \pm 89$ mM. The original three parameters are not much different, apart from having larger estimated errors associated with them, and the new parameter K_{iu} is estimated so imprecisely as to suggest that any uncompetitive component that may be present is too weak to be the cause of the inverted U seen in the first residual plot, an impression confirmed by the fact that the inverted U is still evident when the new fit is shown as a residual plot (not illustrated here, but qualitatively indistinguishable from Fig. 7.1).

As these parameter values were obtained using weights deduced from the data (i.e. weights that can vary during the analysis), the sums of squares in the two cases cannot be directly compared because they have different dimensions. However, the comparison can be done by freezing the weights at the values at the end of one of the two calculation (it does not matter greatly which, as the two fits are almost the same; however, as the conclusion suggested already by the residual plots is that mixed inhibition does not offer a significant improvement over competitive inhibition, it is more appropriate to freeze the weights obtained with competitive inhibition). This recalculation yields trivially different parameter values from those obtained before ($V = 98.1 \pm 10.3$ μmol min^{-1}, $K_m = 2.19 \pm 0.27$ mM, $K_{ic} = 0.312 \pm 0.023$ mM for competitive inhibition; $V = 99.8 \pm 11.8$ μmol min^{-1}, $K_m = 2.25 \pm 0.42$ mM, $K_{ic} = 0.328 \pm 0.0049$ mM, $K_{iu} = 6 \pm 15$ mM for mixed inhibition). In the case of

Table 7.3 Comparing models for inhibition by acetyl-L-alanine

Source	SS	df	MS	F
Total (corrected for V, K_m, K_{ic})	0.40482152	29		
$K_{iu}|V$, K_m, K_{ic}	0.00491237	1	0.00491237	0.344[a]
Residual	0.39990915	28	0.01428247	

[a] Not significant ($F < 1$ is never significant).

competitive inhibition this difference results from the fact that the calculation depends weakly on the initial parameter values and weights; in the case of mixed inhibition it results from the fact that the weights now frozen are not exactly the same as those found with the same model previously.

If the sums of squares are compared by setting up an analysis-of-variance in the same manner as in Table 3.3, the results (Table 7.3) confirm the impression given by the residual plots and the very large standard error associated with K_{iu}: that introduction of this fourth parameter makes essentially no difference to the fit, i.e. that there are no grounds for preferring mixed inhibition over competitive inhibition as a model. With such clear-cut results it is hardly surprising that uncompetitive inhibition gives a very poor fit, with estimated errors greater than 100 per cent for all three parameters, and a much more obvious inverted U in the residual plot.

One cannot regard the analysis as finished at this point, as one ought not to leave the data for acetyl-L-alanine without trying to find an explanation for the inverted U that is perceptible in all the residual plots, which implies some systematic effect that has not been taken account of in any of the models. One simple way to check this is to make residual plots of $a/\hat{v} - a/v$ against the substrate and inhibitor concentrations. The plot with inhibitor concentration as abscissa is shown in Fig. 7.2: the systematic arrangement of points is unmistakeable, and leaves no reasonable doubt that the model fails to account for all of the variation in the rates. Although the evidence for curvature in the arrangement of points is not conclusive, there is certainly a strong suggestion of curvature, with the implication that terms in i^2 are needed in the rate equation.

This example also illustrates clearly the importance of choosing the ordinate variable in a residual plot carefully: if one plots $v - \hat{v}$ instead of $a/\hat{v} - a/v$, the systematic arrangement is still detectable, but with much more difficulty (Fig 7.3). The explanation is that although simple models predict a simple straight-line dependence of $a/\hat{v} - a/v$ on each independent variable (eqn 3.30), the corresponding equation for $v - \hat{v}$ is a rational function, of second degree in both a and i. Nonetheless, residual plots do not always behave as simple analyses may lead us to expect (because one cannot easily predict what effect random errors will have, and because the principal underlying cause of the systematic deviations may be quite

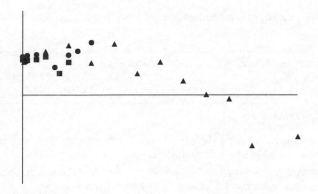

Fig. 7.2.   *Residual plot for inhibition by acetyl-L-alanine.* The plot shows $a/\hat{v} - a/v$, in arbitrary units as a function of i, on a scale that extends from zero to 6 mM. The different symbols represent different substrate concentrations: ■, 0.5 mM; ●, 1.0 mM; ▲, 2 mM.

unforeseen). For this reason, it is usually a good idea to make several residual plots with different combinations of variables.

As noted already, the systematic deviations seen in Fig. 7.2 imply the need for squared or higher-order terms in the rate equation. Conceptually, the simplest way to check this possibility is to fit the data to a model in which the inhibitor can bind to an additional site, thereby generating a term in inhibitor concentration squared in the rate equation:

$$v = \frac{Va}{K_{\mathrm{m}}(1 + i/K_{\mathrm{ic}} + i^2/K_{\mathrm{ic2}}) + a} + e. \qquad (7.4)$$

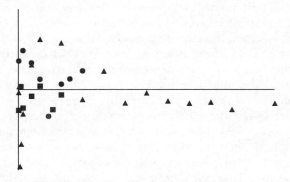

Fig. 7.3.   *Residual plot for inhibition by acetyl-L-alanine.* The ordinate axis shows $v - \hat{v}$; other details are as in Fig. 7.2.

From the practical point of view, however, this is less simple, because eqn 7.4 is not among the equations provided at the outset by Leonora and similar programs. It can be added to the list (see Section 9.3.3), but as adding a new equation is not one of the quickest or simplest functions to carry out, we may prefer to do a 'quick-and-dirty' check for the existence of higher-order behaviour first. This can be done by testing a competitive-inhibition model in which the inhibitor concentration is raised to a power h, i.e. to fit the data to the Hill equation. Leonora does not optimize the value of h automatically, but it does allow the user to define different values from the keyboard with a view to comparing fits by eye (Section 9.5). When this is done, one can quickly determine that values of h around 1.2 give a much more scattered appearance to the original residual plot, with the inverted U largely eliminated, so it seems likely that the binding of acetyl-L-alanine is weakly cooperative.

Adding eqn 7.4 to the list of available equations and fitting it, we find that it fits the data better than the simple competitive-inhibition equation (eqn 7.2), with $V = 103.8 \pm 11.3$ μmol min^{-1}, $K_m = 2.46 \pm 0.30$ mM, $K_{ic} = 0.383 \pm 0.029$ mM, $K_{ic2} = 5.6 \pm 1.3$ mM. Curiously, these parameters correspond to weakly *negative* cooperativity (i.e. to a Hill coefficient less than 1), because K_{ic2}/K_{ic} is much larger than K_{ic}.

One cannot proceed much further than this with the available data. To summarize, therefore, the analysis suggests that the effect of acetyl-L-alanine is more complicated than simple competitive inhibition, and that additional experiments would be needed to establish a better model with certainty.

7.4   Inhibition by acetyl-D-alanine

Application of the same kind of analysis to the data for inhibition by acetyl-D-alanine results in a number of features that may appear puzzling at first sight. The parameter values for competitive inhibition are $V = 47.7 \pm 6.8$ μmol min^{-1}, $K_m = 0.75 \pm 0.18$ mM, $K_{ic} = 4.20 \pm 0.90$ mM, with Michaelis–Menten parameters reasonably similar to those found for acetyl-L-alanine, especially if one notices that the ratio $V/K_m = 63.6$ ml min^{-1} is much closer to the previous value of 44.8 ml min^{-1} than V and K_m are to their previous values. However, introduction of an uncompetitive component to the inhibition gives values for the parameters of mixed inhibition that appear at first sight entirely different from those before: $V = 2450 \pm 9410$ μmol min^{-1}, $K_m = 73 \pm 283$ mM, $K_{ic} = 18.3 \pm 4.3$ mM, $K_{iu} = 0.14 \pm 0.55$ mM. The differences in the inhibition constants present no problem: as the inhibitors are different there is no reason why they should not have different inhibition constants. V and K_m are another matter, however, because the kinetics in the absence of one inhibitor should be exactly the same as in the absence of another (again, the discrepancy for V/K_m is much less than it is for V and K_m individually).

As K_{iu} is much smaller than K_{ic} in this case, the obvious next step is to consider simple uncompetitive inhibition, and now another puzzle emerges, because all the parameter values are negative: $V = -248 \pm 79$ μmol min^{-1}, $K_m = -9.5 \pm 2.5$ mM, $K_{iu} = -1.06 \pm 0.34$ mM (though $V/K_m = 26.1$ ml min^{-1} is again much less aberrant).

The key to all these puzzling features may be found easily by visual inspection of any of the residual plots with $\hat{v}$ as abscissa or, with a little more effort, by studying the values in Table 7.2: one rate, $v = 6.667$ μmol min^{-1} at $a = 2.0$ mM, $i = 0.0$, is more than twice as large as the second largest rate, $v = 3.222$ μmol min^{-1} at $a = 2.0$ mM, $i = 4.0$ mM, and three or more times larger than most of the others. As a result, it exerts a large effect on the estimates of the Michaelis–Menten parameters, especially as it is one of only three observations made in the absence of inhibitor. Looking more closely at these three observations, we see that 6.667 μmol min^{-1} is about 2.45 times greater than the rate of 2.725 μmol min^{-1} observed at half the substrate concentration. In other words, instead of the saturation expected for any enzyme at high substrate concentrations, these two observations suggest an upward curvature in a plot of rate against substrate concentration. Such upward curvature could be an indication of cooperativity, but in the present case there is not enough information to justify such a conclusion, and in any case even a cooperative enzyme normally shows saturation at high concentrations.

It is difficult to escape the conclusion, therefore, that the substrate concentrations do not extend to a high enough value to permit accurate determination of the Michaelis–Menten parameters. All of the K_m estimates obtained in the analysis of the data for acetyl-L-alanine were in the range 2.18–2.46 mM, and even the smallest of these is larger than the largest substrate concentration. The K_m values estimated with acetyl-D-alanine are so disparate that it is hard to draw any conclusion from them, except that enormous standard error estimates are often found with very large K_m estimates (estimates of $1/K_m$ not significantly different from zero); so these values also support the idea that the data do not extend to high enough substrate concentrations. Finally, the analysis above yielded reasonably consistent values of V/K_m but very variable values of V and K_m individually. This behaviour is likewise characteristic of experiments in which the highest substrate concentration is too small.

Returning now to the specific case of inhibition by acetyl-D-alanine, it is clear that we cannot place very much confidence in any of the parameter values calculated, or even in the assessment of the type of inhibition. Nonetheless, there is good reason to believe that it is not acting in the same way as acetyl-L-alanine, as there are obvious qualitative differences in the results, with a clear suggestion of a major uncompetitive component in the inhibition which was completely undetectable for acetyl-L-alanine. This difference is not surprising if one notices the very different inhibitor concentrations needed to produce similar effects in the two cases, some 20 times higher in the case of acetyl-D-alanine.

7.5   Planning future experiments

The analysis given above suggests that the inhibition by acetyl-L-alanine is predominantly competitive, but that there are deviations from simple behaviour sufficiently strong to suggest the possibility of weak cooperativity, or at least of binding at a second site. The data for acetyl-D-alanine suggest that it acts quite differently, not only binding much more weakly but also in a different manner. A possible way to combine these two suggestions would be to postulate that acetyl-L-alanine has two binding sites, of which the less important (weaker) one is the site to which acetyl-D-alanine binds. As noted already, to characterize the enzyme more fully it will be necessary to use a range of substrate concentrations that extends at least to 5 mM, and preferably to about 10 mM. Also, to provide better determination of the Michaelis–Menten parameters one really needs a larger number of observations made in the absence of inhibitor: three is too few for an adequate estimation, and five should be regarded as an absolute minimum. In the present case, increasing the number of observations in the absence of inhibitor from three to ten would hardly affect the characterization of the inhibition behaviour, as it would imply only seven fewer observations in the presence of inhibitor, even if the total number of observations could not be increased (for reasons of the time available for the experiment, for example).

So far as the ranges of inhibitor concentration are concerned, the range for acetyl-L-alanine is adequate if one accepts that simple competitive inhibition is the true model, but it needs to be extended to higher values if one wants to be sure that no uncompetitive component is present, or to establish whether the postulated binding at a second site is real or a statistical artefact. In the case of acetyl-D-alanine, the range of concentrations again appears satisfactory within the limits of the analysis. However, this conclusion may need to be revised after further experiments have been done with a better range of substrate concentrations.

Finally, a suggestion that data analysis are always inclined to make, and experimenters are always inclined to resist: the estimation of parameters and identification of the correct model can in principle always be improved by making more observations. This advice should, of course, be modified to include a proviso that making additional observations should not be accompanied by a decrease in the precision with which they are made. If classical (least-squares) methods of analysis are used, a single grossly inaccurate observation made at the end of a long and tiring day can do much more harm than good to the whole experiment. Even if robust methods are used, as recommended in this book as a general practice, such an observation has little positive value.

The reason why experimenters normally resist such suggestions is that they usually feel, reasonably enough, that they have already made as many observations in each experiment as they reasonably can, and they suspect, often with justice, that whatever number of observations they make the analyst will always suggest that more would have been better. Even suggestions about how a given number of observations are distributed in the design space may be perceived by an experimenter as unrealistic. As one who started his research career working with pepsin substrates that could not be dissolved at higher concentrations than $0.5 K_m$, and even then gave supersaturated solutions that were liable to crystallize during the assay, I am quite conscious that even as simple a suggestion as asking for the substrate concentrations to extend well above K_m may be impossible to satisfy. Even without the question of low solubility, there may be problems of expense, excessive absorbance in the spectrophotometer, etc. Probably none of these applies to acetyl-L-alanine p-nitrophenyl ester, but that is not the point: the point is that in data analysis one is living in an idealized world, but experiments are done in the real world, where not everything that the analysis suggests to be desirable is feasible, and ultimately the experimenter has to be the judge of what it is reasonable to expect.

III Practice

8
Leonora: a program for robust regression of enzyme data

8.1 Introduction

Those interested in the full statistical complexities involved in fitting enzyme kinetic data should consult the review by Garfinkel, which evaluates all these matters thoroughly. However, since the average biochemist will understand none of these articles, we believe it is more helpful simply to present the computer programs and show how to use them.

Cleland 1979

The remainder of this book describes the use of Leonora, a program for fitting steady-state enzyme kinetic equations to experimental data. It is largely self-contained: although the earlier chapters provide an introduction to the theory of the methods used by Leonora, the practical chapters make little direct reference to them. Thus readers who share Cleland's pessimistic assessment of the intellectual capacity of the average biochemist may start at this point, reflecting on the fact that the earlier chapters that they will not read nonetheless make the program less expensive to buy, as they enable it to be sold at book prices rather than computer software prices.

Leonora is intended to allow enzyme kinetic data to be fitted by any reasonable method, including linear and non-linear least squares, robust regression, least absolutes, median regression, minimax fitting, and on-screen graphical estimation. It is aware of all of the common equations—Michaelis–Menten equation, competitive inhibition, two-substrate kinetics, bell-shaped pH profiles, and many others—and allows additional ones to be defined by the user. These must be generalizations of the Michaelis–Menten equation, which means that they must give linear equations when written in double-reciprocal form; they can contain up to five variables, i.e. a dependent variable and up to four independent variables, and up to eight parameters. In addition, it can analyse linear equations, such as the straight line, etc. It accepts various different weighting schemes, but for users who do not want to be concerned with niceties of weighting, it does not require any weights to be defined but tries to estimate the most realistic weighting scheme from the data.

Leonora includes capabilities of programs that were circulated as mainframe FORTRAN programs in the 1970s and 1980s for finding median estimates of Michaelis–Menten parameters (Cornish-Bowden and Eisenthal 1974), and for robust regression of generalizations of the Michaelis–Menten equation (Cornish-Bowden 1985). However, as Leonora goes far beyond these earlier programs it is regarded as version 1 of a new program, rather than an update of older ones.

If your needs are simple, i.e. if you want to fit only the most commonly used equations and you do not want to be concerned with different methods or weighting systems, you will find all the essential information for installing and running Leonora in this chapter. The only qualification to this is that if you find the screen colours disagreeable or difficult to read (for example, if you cannot easily tell which characters are highlighted), you may need to refer to Section 9.8.

8.2 Typographical conventions

Various typefaces are used in this guide to distinguish between different sorts of text. Things for you to type into the computer are in a *sloped typewriter style* and results displayed on the screen are shown in the same style, but upright. Menu entry lines are shown sans-serif with key letters that would be highlighted on the PC screen **bold**. Keys for you to press to activate menu options are likewise shown as **bold sans-serif**, apart from words that represent single keys, such as ENTER, TAB, F1, etc., which are shown in SMALL CAPITALS. In using a menu, you can select an entry either by pressing the highlighted key (capital or lower-case) or using the arrow keys ↑ ↓ to bring the triangle ▶ level with the required line and then pressing ENTER. Although capital and lower-case keys are in general equivalent, pressing the upper-case key will execute immediately in cases (such as deleting a file or quitting Leonora) where the lower-case key would require confirmation. If you find that a command unexpectedly fails to demand confirmation, check that the SHIFT-LOCK key is not down. On entry to a menu the ▶ will normally be placed by the entry that Leonora considers most appropriate to choose. You will often find that some of the entries in a menu are unavailable and shown in square brackets: if you try to select such an entry you will obtain an error message indicating why the entry is unavailable. For example, if you try to select [Equation] in the MAIN MENU when starting Leonora, you will obtain the message Data must be entered first. In any menu you can obtain context-sensitive help by pressing F1.

In addition to F1, Leonora also recognizes the function keys F2, F3, and F4. These are described in Section 8.10.

8.3 Installation on a hard disk

To install Leonora on a hard disk, put the 3.5 inch disk in the appropriate drive, and type *A:* ENTER or *B:* ENTER (as appropriate), and then Install ENTER at the DOS prompt if you want to install Leonora on drive C, or *Install D:* ENTER (where D is the appropriate drive letter), if you want to install it on a drive other than C. (In the latter case drive C must still exist, and must still have at least 1 kilobyte of space on it for temporary files.) This will bring up a window suggesting directories for storing the files. You can edit the name C:\LEONORA for the executable file to whatever you wish. The names of the sub-directories for the data files and help files can likewise be edited, but these must be sub-directories of the executable directory. The sub-directory for private files is shown for information, but it cannot be edited. When you are ready, press ENTER. All of the files will be copied into the appropriate directories, and the screen will make a suggestion for modifying your AUTOEXEC.BAT file. However, the installation program will not itself make any alteration of this file (because most users prefer to deal with their AUTOEXEC.BAT files themselves).

If for any reason you interrupt the installation before it is finished (or if your computer crashes while installation is in progress), you should then type *del c:\leondirs.bat* before trying again to install. (In the normal installation this file is created during the first stage and deleted at the end; installation is unlikely to proceed normally if it exists at the outset.)

Leonora is then ready for use. If you have followed the suggestion for modifying your AUTOEXEC.BAT file you can run Leonora by typing *LEONORA* ENTER at the DOS prompt in any directory. If not, you must change to the appropriate directory, by typing *CD C:\LEONORA* ENTER and then *LEONORA* ENTER.

To remove Leonora from your hard disk, type *LEONORA UNINSTALL* ENTER, or *LEONORA UNINSTALL FORCING* ENTER. The parameter *FORCING* causes removal of the files to proceed without any pauses to allow you to check each file before it is deleted.

8.4 Example 1: the Michaelis–Menten equation

This and the following sections illustrate the use of Leonora to fit the equations most commonly encountered: the Michaelis–Menten equation for substrate/rate data, competitive inhibition for substrate/inhibitor/rate data, ternary-complex equation for substrate/substrate/rate data, and bell-shaped pH profile for pH/parameter data. A brief account of fitting other kinds of equations to the same data will be found in Section 8.8, with a more detailed treatment in Section 9.3.

Fitting the Michaelis–Menten equation will be illustrated by reference to problem 12.1 from *Fundamentals of enzyme kinetics* (Cornish-Bowden 1995). We shall enter and analyse the following set of ten ([S], v) pairs, where each v is the initial rate at the corresponding substrate concentration [S]:* (1, 0.219), (2, 0.343), (3, 0.411), (4, 0.470), (5, 0.490), (6, 0.525), (7, 0.512), (8, 0.535), (9, 0.525), (10, 0.540).

Launch Leonora by entering its name *LEONORA* at the DOS prompt. The opening welcome screen will appear and after a few seconds (or immediately, if you press any key) the MAIN MENU will appear. This will show three active options **Last data**, **Data**, and **Exit**, and four inactive options shown in square brackets, [**Equation**], [**Output requirements**], [**Calculations**], and [**Plot results**]. Press **D** to bring up the DATA MENU, and then press **I** to activate the option **Input new data**. Type *[S]* as the caption for column 1, press TAB to move to column 2, and type *v* as caption for this column. Although Leonora allows you to have up to five columns of data, we need only two for this example.

Next press ENTER to reach column 1 of line 2 and type *1* as the first [S] value. You can enter the values in any order you like, but when the [S] values follow a logical progression, as in this case, it is usually simplest and quickest to enter them all before starting to enter the v values. To do this, press ENTER to reach the next line, type *2*, then press ENTER again, type *3* and continue in this way until you have typed *10* for the last [S] value.

Then use the arrow key ↑ to reach the line for [S] = 1, press TAB to move to column 2, and then type *0.219* and press ↓ to reach the next line. After typing *0.540* as the last v value, press any arrow key or ENTER and then press ESCAPE. (Do not press ESCAPE while still in the cell containing the value *0.540*, because if you do Leonora interprets this as meaning that you do not want a value in this cell.)

Some points to note are the following.

1. If you make an error while entering a value, you can correct it by using the arrow keys to return to the cell with the error. If you only discover the error after the whole set of data has been entered you can still correct it by using the **Edit data** option in the DATA MENU. (An example is described in Section 8.5.)

2. It is not illegal to enter non-numerical values in a cell where a number would normally be expected, but lines containing such values are ignored during the calculation. You may do this deliberately as a way

---

* Notice that although elsewhere in this book the substrate concentration has usually been represented by the symbol a, Leonora has no preconceptions about the symbols for different variables and allows considerable flexibility in the choice. Here we shall follow a different convention and use [S].

of removing certain observations from the calculation while leaving it easy to reinstate them later. For example, you might replace 0.411 by *(0.411)* to do a calculation without the third observation. As a warning against doing this by accident (for example after typing the letter *O* in place of the number *0*), values that cannot be interpreted as numbers are displayed in warning style.

3. The TAB key always moves you to the right. If you want to move left, press ENTER to get to column 1 of the next line, or SHIFT-TAB (i.e. press TAB while holding the SHIFT key down) to move one column left in the same line.

4. Although the values can be entered in any order, you cannot move to a column that has no caption. Thus if you have defined two captions *[S]* and *v* as suggested above, pressing TAB while in column 2 will not bring you to column 3 unless you are in line 1.

5. After you enter a value and move to another cell, the value you typed is displayed as interpreted by Leonora: for example, 9 will be displayed as 9.0000.

As soon as you press ESCAPE a window will appear in which you will be asked to define what kind of data is in each column. (As you can arrange the columns in whatever order you like, Leonora has no preconceptions about what will be in each column.) Press **S** to define [S] as a substrate concentration, then press **D** to define *v* as the dependent variable. (As the dependent variable will often have the symbol *v*, Leonora accepts **V** as synonymous with **D**.) If you have made no mistake, press ENTER to register the information, ESCAPE to return to the MAIN MENU without defining a new set of data, or any other key if you want to correct the column definitions.

After you press ENTER you will return to the DATA MENU where you should press **T** or ENTER to choose the option **Title data**. (This step is not compulsory, but it helps to overcome the problem that 8-character DOS names are too short to allow much information about how one data set differs from another.) Then type the title *Problem 12.1 from 'Fundamentals of Enzyme Kinetics'* and press ENTER.

Next press **N** or ENTER to choose the option **Name and save data**. Type *EXAMPLE* in the window that appears, and press ENTER. Then press ENTER or **X** for **Exit** to the MAIN MENU. The data set is actually the same as the one already stored as EXAMPLE1.MMD when you install Leonora; however, use a different name here to avoid over-writing it.

Leonora knows that with ([S], *v*) pairs as data the most likely equation for you to want to fit is the Michaelis–Menten equation, so at this point you can ignore the **Equation** entry in the MAIN MENU and proceed directly to CALCULATIONS by pressing **C** or ENTER. (We shall also ignore the **Output requirements** option at this stage.)

Table 8.1   Results from example 1

(a) Parameter screen.

```
              RESULTS FROM WHOLE SET OF DATA
           Determined by least squares (by default)
                Converged after 5 iterations
        Weights calculated with adjustable σ/σ (by default)
                 Km/V = 2.82206 ± 0.16839
                 1/V = 1.51579 ± 0.03948

                   V = 0.65972 ± 0.01718
                  Km = 1.86178 ± 0.15339

                 v = V·[S]/(Km + [S])
```

(b) Calculated rates, errors, and weights.

	RESULTS FROM WHOLE SET OF DATA				
[S]	v	v*	Error	Wt1	Wt2
1.00000	0.21900	0.23053	-0.01153	2.3788	0.87
2.00000	0.34300	0.34167	0.00133	1.5405	1.00
3.00000	0.41100	0.40709	0.00391	1.2040	0.99
4.00000	0.47000	0.45019	0.01981	1.0591	0.83
5.00000	0.49000	0.48072	0.00928	0.9366	0.97
6.00000	0.52500	0.50349	0.02151	0.8850	0.83
7.00000	0.51200	0.52112	-0.00912	0.7961	0.97
8.00000	0.53500	0.53518	-1.76e-04	0.7726	1.00
9.00000	0.52500	0.54664	-0.02164	0.7206	0.86
10.00000	0.54000	0.55618	-0.01618	0.7062	0.92

$$v = V·[S]/(Km + [S])$$

The CALCULATIONS MENU contains several options, but we shall ignore all of these except the first, in effect allowing Leonora to choose the method of fitting and system of weighting. So press **C** or ENTER to start the calculation. When the calculation is complete the parameter values will be displayed on the screen as illustrated in Table 8.1(a).

The top lines of the window show what method and weights have been used. The two parentheses (by default) mean that both method and weights have been assigned by Leonora and not by calling the METHODS (Section 9.5.1) or WEIGHTING SYSTEM (Section 9.5.2) menus from the CALCULATIONS MENU (Section 9.5). The equation that has been fitted appears at the bottom of the window. When written in double-reciprocal form as the expression for a straight line, the Michaelis–Menten equation $v = V[S]/(K_m + [S])$ becomes $1/v = 1/V + (K_m/V)/[S]$. The two parameters $1/V$ and K_m/V that appear in this transformation are those used in the actual calculation and are listed first in the results. After them are shown the parameters V and K_m used in the definition of the equation.

When any key is pressed a new window appears showing six columns of results, as shown in Table 8.1(b). The first two columns just reproduce the original data: these should be checked for mistakes. The column headed v* shows the values of v calculated with the parameter values that have been estimated. However, you should not take my word for it—all programs contain bugs, and much as I should like to believe that Leonora is the exception, it probably is not, so results that can easily be checked should be checked. Using a pocket calculator, one can calculate $0.65972 \times 8/(1.86178 + 8)$ as 0.535173163, which agrees satisfactorily with the value shown on the screen. The column headed Error gives values of v - v*, for example $0.41100 - 0.40709 = 0.00391$. If you are of a suspicious frame of mind you should notice that the sequence of signs in this column $(- + + + + + - - -)$ does not look very random and should consider how it might be explained, but for the moment we will ignore this. The next two columns Wt1 and Wt2 show the heteroscedastic and robust weights respectively that have been found, but as this section is written with the assumption that you are not interested in weighting, we will ignore these columns.

When you press any key a residual plot will occupy the whole screen: each value of the weighted difference between v and v* is plotted against v*. To allow even very deviant observations to be plotted without squashing all the others against the horizontal axis, the ordinate scale is not linear but is a trigonometric transformation of the actual weighted differences which has a negligible effect on the smaller deviations but decreases the magnitudes of the largest ones. The horizontal axis is also shown, but otherwise the plot is completely unlabelled so as not to decrease the visual impact of the plot, which you should study in relation to two questions.

1. Do the points appear to be non-randomly scattered, with an obvious trend? In this example the answer should be *yes*, as the points should lie approximately on an inverted U: this indicates that the Michaelis–Menten equation is probably not the most appropriate equation for the data, a point we shall return to later.

2. Are any points very obviously out of agreement with the others? In this example the answer should be *no* and we do not need to consider the question further.

Pressing a key causes each point in the residual plot to be labelled with its v value: this is to enable you to identify easily any points of interest. For example, if all the points except one were placed very close to the axis, you would expect the exception to be the result of a typing error, and you would want to know which value to correct.

Pressing a key now returns you to the CALCULATIONS MENU. If you want to see the parameter values or results again, press **P** or **R**, but otherwise you can press **X** or ENTER to return to the MAIN MENU, as the calculation for this example is finished.

If you now press **P** the PLOTTING MENU will appear, and you should press **A** or ENTER to select the option **Axes**. On entry to the menu that appears the axes will be shown as [S] for the abscissa and v for the ordinate. However, you can change these with the arrow keys. For example, to obtain a plot of [S]/v against [S], you must leave the abscissa as it is, and modify the ordinate as follows. First press TAB to move the triangle ▶ to the line labelled Ordinate; then press ↑ to change the entry to [S]*v; then press → to put the triangle ▼ over the column labelled v; finally press ↓ twice to change the entry first to [S] and then to [S]/v. At this point the message LINEAR appears on the bottom line, indicating that the choice of axes is one that gives a straight line plot in observation space. The axes are now as required, so press ENTER.

A number of comments about plotting with Leonora can be made.

1. Although Leonora can draw either curves or straight lines, depending on the axes chosen, a curve takes much longer to draw than a straight line. This effect may be quite noticeable if your computer is slow.

2. When you first opened the window for changing the axes the message DIRECT PLOT POSSIBLE was visible, but disappeared when the axes were no longer shown as [S] and v. This message indicates that for the choice of axes shown the locus of parameter values satisfying any observation exactly is a straight line in parameter space. If you make such a choice the option **Direct linear plot** will become active when you return to the PLOTTING MENU.

3. Leonora can make a plot with any combination of variables with each to the power −1, 0, or 1. Not all combinations make biochemical sense, however. For example, if you plot 1/[S] against [S] you will obtain some information about coordinate geometry but none about biochemistry or about your data.

On returning to the PLOTTING MENU, press **S** or ENTER to examine the scale ranges. This step is not compulsory, but if you omit it the plot will extend over the range of the data and no more, which is unlikely to be what you want. Press **Z** twice to make both axes extend as far as zero. Then press TAB to move the cursor to the right-hand column and edit 10.900 to read *12*; press ↓ to reach the line labelled Ordinate and edit 19.914 to read *20* or *25*. Then press ENTER.

On returning to the PLOTTING MENU, select **Plot** by pressing **P** or ENTER. The data will then be plotted, the line being calculated with the last parameter values obtained.

If your computer is short of memory (e.g. because you run Leonora with another program resident in memory) you may get a warning after pressing **P**. When this happens Leonora cannot store the font used for plotting in memory, but has to read it as required from a disk file. Whether this is a problem or not will depend on how fast your computer is.

The menu that appears at the bottom of the screen allows you to see the effects of changing the parameter values. Initially Km/V is selected, but you can toggle between Km/V and 1/V using TAB. Pressing ↑ causes the selected parameter to be increased by 1 per cent; pressing ↓ causes it to be decreased by 1 per cent. You can increase or decrease the amount of change in each step by pressing PAGE-UP or PAGE-DOWN.

You can test your skill at fitting data by eye on a graph by varying the parameters in the way just described until you think a best fit has been achieved. For doing this it is best to press **Y**, for Fit by eye before pressing **P**. This has the effect of altering each stored parameter value by a small random amount, so that you cannot be biassed by knowing what line the calculated parameters give.

You can return to the PLOTTING MENU by pressing **X**, from where you can either make additional plots with different axes, or you can press **X** a second time to return to the MAIN MENU. You can also make a residual plot instead of a direct one by selecting **Residual plot**. You will recall that the residual plot that appeared automatically after the results were displayed suggested some systematic variation: if this were the consequence of substrate inhibition, you might expect the dependence of residual on substrate concentration to be more evident than the dependence on rate. To check this, return to **Axes** to check that the abscissa is still [S] and the ordinate is still v, and then select **Residual plot**. This will then generate a plot of Residual (v), i.e. $v - v^*$, against [S], in which the systematic tendency will be very obvious.

After pressing **X** to return to the MAIN MENU, you can press **X** again to exit from Leonora or you can proceed to another example, such as the one in Section 8.5. Better still, you can check the suggestion that the data might fit the equation for substrate inhibition better than the Michaelis–Menten equation by choosing this equation from the EQUATIONS MENU and repeating the calculations and plots. This is discussed in Section 8.8.

8.5   Example 2: competitive inhibition

For an example of fitting inhibition results we take the data of problem

Table 8.2   Parameter values for example 2

The upper panel shows results for the data as typed in originally, i.e. 15 ([S], [I], v) triplets as follows: (1, 0, 2.36), (2, 0, 3.90), (3, 0, 5.30), (1, 1, 1.99), (2, 1, 3.35), (3, 1, 4.40), (1, 2, 1.75), (2, 2, 2.96), (3, 2, 3.98), (1, 3, 1.60), (2, 3, 2.66), (3, 3, 8.53), (1,4, 1.37), (2, 4, 2.35), (3, 4, 3.33). The lower panel shows the effect of correcting (3, 3, 8.53) to (3, 3, 3.58).

$$Km/V = 0.33548 \pm 0.00667$$
$$Km/(V.Kic) = 0.07692 \pm 0.00198$$
$$1/V = 0.08952 \pm 0.00316$$

$$V = 11.17115 \pm 0.39374$$
$$Km = 3.74767 \pm 0.15165$$
$$Kic = 4.36124 \pm 0.14163$$

$$Km/V = 0.33503 \pm 0.00621$$
$$Km/(V.Kic) = 0.07725 \pm 0.00181$$
$$1/V = 0.08967 \pm 0.00295$$

$$V = 11.15251 \pm 0.36699$$
$$Km = 3.73639 \pm 0.14109$$
$$Kic = 4.33718 \pm 0.12936$$

5.3 of *Fundamentals of enzyme kinetics* (Cornish-Bowden 1995) which consist of 15 ([S], [I], v) triplets as follows: (1, 0, 2.36), (2, 0, 3.90), (3, 0, 5.30), (1, 1, 1.99), (2, 1, 3.35), (3, 1, 4.40), (1, 2, 1.75), (2, 2, 2.96), (3, 2, 3.98), (1, 3, 1.60), (2, 3, 2.66), (3, 3, 8.53), (1, 4, 1.37), (2, 4, 2.35), (3, 4, 3.33).*

First select **Data** from the MAIN MENU, followed by **I**nput new data from the DATA MENU, exactly as above. Then enter the observations in the same way as before, with the difference that now there are three columns of data instead of two. When prompted to define the kind of data in each column, press **S** and **D** (or **V**) for the columns [S] and v, as before, and **I** for the new column, which should have been labelled [I]. Proceed as before to the CALCULATIONS MENU, again ignoring the **E**quation entry in the MAIN MENU, because in the absence of other information Leonora assumes that you will want to fit the equation for competitive inhibition to ([S], [I], v) data. After pressing **C** or ENTER to start the calculation, you will obtain the results shown in the upper panel of Table 8.2.

Notice that in the screen of results obtained by pressing any key, the entry in the Error column for the observation (3, 3, 8.53) is more than ten times larger than any other value in the column. In a less extreme example, or with a larger data set, this might pass unnoticed in the results screen. It should be quite obvious, however, in the residual plot obtained

---

* Although the supplied file EXAMPLE2.MMD may appear at first sight to contain these data, it is actually slightly different and will not give the same results as those described here.

by pressing a key that the axis is very far from the middle of the screen and that one observation is very far from the axis. Pressing a key again reveals that this observation has $v = 8.53$. Reference to the source of the problem shows that this value should have been 3.58, and so it needs to be corrected.

To correct a mistake in the data, press a key to return to the CALCULA-TIONS MENU, press **X** to exit from this menu to the MAIN MENU, and **D** to open the DATA MENU. There press **E** for the option **E**dit data. This will open a window containing the data. Press ↓ repeatedly until the highlighted cell reaches the offending line; then press → twice to bring it to the right-hand column and press **E** (or ENTER) to start editing. Then edit 8.53 to become 3.58 and ENTER. Finally press **X** to exit from the editor, press **V** to save the edited data, and **X** to return to the MAIN MENU.

Repeating the calculation should now give the results shown in the lower panel of Table 8.2. The precision estimates are smaller, though not as much smaller as you might expect from the grossness of the original error, and the parameter estimates themselves are very little changed. This is because if nothing else is specified Leonora uses a 'robust' method of fitting that allows it to ignore observations that are grossly aberrant. Examination of the detailed results and the residual plot now reveals no very obvious anomalies, though there is a weak suggestion of a U-shaped distribution of points, and you might feel it worthwhile to check the point $(3, 0, 5.30)$ for a possible typing error.

8.6   Two-substrate kinetics

Data for a reaction of two substrates can be dealt with in essentially the same way as inhibition data, and thus do not require a detailed example. The main difference is that in defining the data type for each column you must respond **S** for both of the columns containing substrate concentrations. With data of this type Leonora assumes that you want to fit the equation for a ternary-complex mechanism.

8.7   Example 3: pH-dependence data

For illustrating how one might fit a bell-shaped pH profile, consider the following series of measurements of k_{cat}/K_m as a function of pH for the pepsin-catalysed hydrolysis of acetyl-L-phenylalanyl-L-phenylalanylglycine, each observation being a (pH, k_{cat}/K_m) pair: (0.47, 37), (0.72, 74), (1.12, 176), (1.67, 237), (1.80, 192), (2.08, 292), (2.30, 198), (2.57, 233), (2.75, 215), (3.28, 164), (3.98, 70.6), (4.59, 23), (4.79, 13.2), (5.20, 3.5), (5.67, 0.98). These can be entered in the same way as in Example 1, except that the columns should be labelled *pH* and *kcat/Km*, and you should respond **P** and **D** respectively to the request to show the data types of the two columns. With data of this kind Leonora assumes that a bell-shaped curve is to be fitted, and calculation with the pepsin data produces $k_{\lim} = 247 \pm 42$, $K_1 = 0.069 \pm 0.040$, $K_2 = (2.6 \pm 1.3) \times 10^{-4}$ as parameters of the equation $k_{cat}/K_m = k_{\lim}/(1 + [H^+]/K_1 + K_2/[H^+])$.

8.8   Fitting other equations

To this point we have mainly discussed data fitting as if only one kind of equation were appropriate for any data set. In reality this is not the case, and one will often want to compare the fits obtained with different equations. For each kind of data Leonora knows of several appropriate equations, and it is possible to add others.

In fitting the Michaelis–Menten equation to the data of Example 1 (Section 8.4), there was a non-random distribution of points in the residual plot suggesting that it might be appropriate to try a more complex model. In this case the most obvious possibility is to test for substrate inhibition. To do this, it is first necessary to recover the data by choosing Data from the MAIN MENU, and then Find data file in the DATA MENU. This produces a window showing all of the data files in the data directory. Press ↑ until the file EXAMPLE1.MMD is shown against a bright background; press ENTER to select this file. (It will be read faster if you press ENTER again while it is being read.) When the file is read press ENTER again to return to the MAIN MENU and then select Equation, which will produce a list of equations that can be fitted to the data. These will be listed by name, but if you prefer to see the actual equations you can achieve this by selecting Show equations algebraically. In either case, press S to select the equation for Substrate inhibition, and then repeat the calculation as before.

When you do this, you will find the fit may appear much better to the eye, but Leonora will give you no numerical information on which to base a more objective assessment, and the entry [Show statistics] will be shown in brackets in the CALCULATIONS MENU. This is because we have (by default) used dynamic weights, i.e. we have allowed Leonora to select the best weighting scheme, and comparison of sums of squares between models is meaningless if these have been obtained with different weighting schemes. To overcome this, select Define weighting system from the CALCULATIONS MENU, then Fix current weights and Exit from the WEIGHT-ING MENU. At this point you may notice that the option Show statistics, which was previously given in square brackets, is now available. If you select it by pressing S, you will obtain the following statistical information on successive lines: the number of degrees of freedom, the 'effective number of degrees of freedom', i.e. the sum of robust weights minus the number of parameters, the mean square residual, its square root, and the sum of squares scaled up to reflect the fact that the robust weights add up to less than the number of observations. This last value is the sum of

Table 8.3   Sum of squares calculation

The table shows the way in which Leonora calculates and displays the sum of squares. The items are labelled with algebraic expressions rather than symbols to avoid any misunderstanding of the definitions.

$$\Sigma \text{w1.w2}(\text{v} - \text{v*})^2 = 3.616\text{e-}04$$
$$\text{n} - \text{p} = 7$$
$$\Sigma \text{w2} - \text{p} = 6.2740$$
$$\Sigma \text{w1.w2}(\text{v} - \text{v*})^2/(\Sigma \text{w2} - \text{p}) = 5.764\text{e-}05$$
$$\sqrt{[\Sigma \text{w1.w2}(\text{v} - \text{v*})^2/(\Sigma \text{w2} - \text{p})]} = 0.00759$$
$$(\text{n} - \text{p}).\Sigma \text{w1.w2}(\text{v} - \text{v*})^2/(\Sigma \text{w2} - \text{p}) = 4.035\text{e-}04$$

squares used in significance tests. To make it clear exactly what has been calculated, these values are shown on the screen in algebraic form rather than by name, as shown in Table 8.3. There is also a last line indicating that as no replicate observations were found in the data no analysis of pure error could be done.

Before trying to make sense of these numbers, you should repeat the calculation for the Michaelis–Menten equation. To do this, return to the EQUATIONS MENU, select Michaelis–Menten again, return to the CALCULA-TIONS MENU and repeat the calculation. The parameter values ($V = 0.66540$, $Km = 1.92000$) are not exactly the same as those obtained originally (because we have changed the weighting), but the important change is that now two statistical screens appear automatically after the results screen. The first contains new values for the quantities detailed for the substrate-inhibition results, but the last value, the scaled up sum of squares, is now much larger, 0.00185 instead of $4.035\text{e-}04$. The second statistical screen displays an analysis of variance, showing that the additional param-eter has decreased the sum of squares by 0.00145103, with an F value of 25.173 for 1 and 7 degrees of freedom. The estimate of $P = 0.0015$ indicates that it is a highly significant improvement.

In a similar way you can repeat the calculation for the data of Example 2 with the equations for uncompetitive and mixed inhibition. In this case you should find that uncompetitive inhibition gives a definitely poorer fit than competitive, and that mixed inhibition reveals hardly any improve-ment over competitive. Although it might be tempting to conclude from this that the competitive equation is definitely the correct one for the data set, a more critical approach would be to ask whether the experiment was well designed to yield information about any uncompetitive component that might be present.

A fuller account of the EQUATIONS MENU is given in Section 9.3.

8.9   Screen layout

When you first launch Leonora and arrive at the MAIN MENU, the top line should show the words Leonora Michaelis-Menten at the left, and the date and time at the right. As long as the colon : in the time is blinking, the time is being continuously updated. However, it is not updated when Leonora is not expecting a menu selection to be made (e.g. while you are entering or editing a numerical value). At such times the colon either stops blinking or disappears, and the cursor is normally visible at the active location of the screen. If you have entered a new set of data or edited an old one, but not saved it, an asterisk * is shown at the centre of the top line. Once an equation has been selected it is shown (often truncated) at the top-left of the screen.

On first launching Leonora the bottom line of the screen is blank, unless you have modified the default setting for automatic help (Section 9.8), but as soon as a set of data has been entered or read, its file name and title are shown on this line. This line then only becomes blank again after you delete a data file or during the time that a plot is on the screen. If you prefer to use this line for displaying help information, press ?.

The rest of the screen is used for menus or results. Menus normally grow from the bottom-right of the screen in such a way that the current menu is completely visible and partially obscures its parent menu. Deviations from this arrangement occur when defining a new set of data or editing an existing one, because in these circumstances it is necessary to have both results and menus or other information on the screen at the same moment.

8.10   Miscellaneous points

Memory management. To check how much memory is free at any moment, you must first prime Leonora by pressing \ while at the MAIN MENU. Subsequently pressing μ (ALT-230) at any menu will cause two lines to appear giving the values in bytes of the Turbo Pascal variables MemAvail (total available memory) and MaxAvail (largest single block of memory). If these are always greater than 30 000 you are unlikely to have any memory problems. Press any key to restore the menu to its normal state.

In general Leonora tries to reserve memory for the more important activities, with the result that certain cosmetic functions may be omitted when memory is low. In such cases the message Not enough memory to restore the screen correctly will be displayed, and the menu that follows may be incorrectly formatted or in the wrong part of the screen. (This message may sometimes appear when there is no fault, because Leonora normally makes somewhat pessimistic estimates of how much memory is needed.)

Special function keys. In addition to F1 (help) and ALT-F9 (sudden death, Section 9.1), Leonora recognizes the function keys F2, F3, and F4. Pressing F2 at almost any moment allows you to write notes in a file called

NOTES.TXT without quitting Leonora. If this file already exists it will be displayed first, but if you press a key while it is being displayed you can then either delete the file by pressing DELETE, or start adding to it by pressing any other key.

Pressing F3 has no effect except to display the message BROKEN IN: Press any alphabetic key. If you then press an alphabetic key execution will resume as before. (If you press a non-alphabetic key execution will resume but the top line of the screen will remain blank.) This key was used in writing and debugging Leonora and has no function in ordinary use.

Pressing F4 while a menu is on the screen will cause the number of lines of the file LEONORA.MNU that were read before the correct menu was found, and the time in milliseconds required to do this and to construct the menu. This may be helpful for knowing whether you are likely to be able to decrease the time taken to display frequently used menus by bringing them nearer to the beginning of the file.

Why is the program called Leonora? If you look at the original paper of Michaelis and Menten (1913) you will find the authors listed as 'L. Michaelis and Miß Maud L. Menten': the first L. stands for Leonor, the second for Leonora. The strange word Miß, which may be difficult to find in a German–English dictionary, is best read as an English word printed as if it were German.

9
Leonora menus

9.1 Main menu

This chapter lists the menus that Leonora contains, beginning with the MAIN MENU, both for reference and to give a more complete description of the program than was given in Chapter 8. Remember, however, that Leonora can be very extensively customized (Chapter 10) and so although the information here corresponds to the program as it appears immediately after installation, it may not agree with the version you are actually running.

The MAIN MENU contains the following entries.

1. **Last data: FILENAME.MMD.** This entry is for selecting the same set of data as at the last use of Leonora without having to search for it in the data directory. If this set of data is already read, or if the data file has been moved from the data directory since the last time Leonora was run, the entry appears as [Last data: FILENAME.MMD]; if the file name has been lost it appears as Last data (unknown).

2. **Data.** This entry gives access to the DATA MENU, which contains entries for reading or editing existing files, or defining new ones (Section 9.2).

3. **Equation.** This is for defining the equation to be fitted (Section 9.3).

4. **Output requirements.** This is for specifying whether you want an output file (for subsequent printing or further analysis), and if so whether you want it to contain just results or details of the calculation as well (Section 9.4).

5. **Calculations.** This is for carrying out the calculations, and also for selecting a method and defining weights (Section 9.5).

6. **Plot results.** Provided that your computer has a graphical capability (now nearly always the case) and sufficient memory is free for loading the graphics routines, this allows you to plot the results of the calculation on screen (Section 9.6).

7. **Exit program.** This allows you to quit from Leonora, after pressing ENTER to confirm that this is what you want to do. A warning message

will appear if you try to quit when the current set of data is not saved. If you have enabled the ESCAPE key for escaping from the MAIN MENU (Section 9.8) this entry appears as Exit (or escape) program. Regardless of whether the ESCAPE key is enabled, you can escape from Leonora at any menu by pressing ALT-F9: use this option with care—it provides no opportunity to change your mind and does not check that the current set of data is saved.

9.2   Data menu

The function of the DATA MENU is to allow data files to be written, read, altered, or deleted. It contains the following entries.

1. Find data file. This command generates a list of files in the current data directory with the extension MMD (= 'Michaelis–Menten data'). Pressing TAB converts this to a list of *all* files in the data directory, but many of these will be unreadable or will generate an error if you try to read them. However, back-up files (MMK) and 'hidden' files (MMH) should normally be usable. You can choose a file by pressing the appropriate arrow keys until the file you want is shown against a bright background, and then pressing ENTER. If you have data stored in more than one directory (which must be sub-directories of the directory in which the program is stored) you can change directory by selecting the entry at the top of the list, Change Directory. If no additional directories have been created with DOS, this will show only the current directory, but if others exist they can be selected in the same way as selecting files. Leonora itself contains no mechanism for creating directories or transferring files from one directory to another.

2. Input new data. This is for defining a new set of data, as described in Section 8.4 (for two variables) and Section 8.5 (for three variables). The screen consists of three regions, a data window occupying most of the top-left of the screen, help information in the lower part, and a column of reference information at the right:

   (a) *Data window.* The data window allows up to five columns of data and a number of rows limited only by the amount of memory. The top row is for labels (symbols for the variables); the others are for values. Values can be entered in any order, except that you cannot move to the right from a completely empty column, for example you cannot enter data in column 3 if column 2 is empty, and (apart from column 1) you must assign a label to a column before entering data in it. Non-numerical values can be entered in value cells but will be ignored in the calculations.

(b) *Help region.* As well as indicating the numbers of rows and columns allowed, this defines the effects of the following keys: TAB, move one column to the right; SHIFT-TAB, move one column to the left; ENTER, go to column 1 of the next line; PAGE-DOWN, move one row down, copying all values to the left into the next row (this is useful for entering replicate or near-replicate observations); PAGE-UP (not shown on the screen), move to column 2 of the next row, copying the value in column 1 of the current row; ESCAPE, terminate (return to the data window). The following keys are not listed on the screen and have their ordinary editing functions in the cell that contains the cursor: →, ←, HOME, END, INSERT, DELETE. The arrow keys ↑ and ↓ can be used to move from one row to another.

(c) *Information region.* The top line gives the amount of memory currently available. Divide this by 32 to estimate how many cells can be filled. Unless you have other programs resident in memory while running Leonora the memory will normally be much more than you are likely to need. The next three rows indicate how many cells, rows, and columns respectively contain values or labels. Once you have entered at least one value, the next two lines indicate the last value entered.

3. Save data file. Use this command to save to disk the new (edited) version of an existing file. The previous version is *not* deleted but is renamed with the extension MMK. Any existing MMK file with the same name is lost. You can use the same command for a newly created set of data, but if you do the file will be given the name NEWDATA and the title Untitled data. It is better, therefore, to use **Name and save data** (item 6 below).

4. Hide data. This command renames the current set of data to have the extension MMH. The file still exists but no longer appears in the list created by **Find data file** (item 1 above) unless TAB is pressed.

5. Delete data. This command is for deleting the file containing the current set of data. If the current data have not been saved, no file exists and thus it cannot be deleted. If a file does exist, the command must be confirmed before it is executed, unless you force execution by pressing **D** (SHIFT-D) rather than **d**.

6. Name and save data. This is for saving the current set of data under a new file name, either because it is a new set of data or because you want to retain the original file as well as the new one.

7. Title data. This is for giving a descriptive title (up to 60 characters) to the current set of data.

8. **Edit data.** This is for editing the current set of data and is described in Section 9.2.1.

9. **Redefine data type.** If you have made an error while defining the kind of data after entering a new set of data (Sections 8.4–5), you can use this command to correct it.

10. **Default directory: DIRNAME.** If you have read a data file from a directory DIRNAME that is not the current default directory, you can use this command to redefine the default so that in future Leonora will look first in DIRNAME.

11. **Exit.** Return to the MAIN MENU.

9.2.1 Editing menu

The EDITING MENU differs from other Leonora menus in that it is active at the same time as a window of data is visible. Its entries can only be chosen with the highlighted keys, not with the arrow keys, because these are needed for navigating the data window. The cell that will be acted on by any command is shown against a bright background. The menu contains the following entries.

1. **Delete observation.** If confirmed by pressing ENTER, this command causes the whole of the line containing the highlighted cell to be deleted. If you use the SHIFT key, i.e. press SHIFT-D, to force the selection the line will be deleted without confirmation. The top line (containing names for the columns) cannot be deleted.

2. **Duplicate observation.** This command causes the whole of the line containing the highlighted cell to be duplicated.

3. **Copy cell to new row.** A new row will be created in which the highlighted cell is copied and the other cells are undefined (indicated by ?). The highlighted selection moves to the first undefined cell.

4. **Edit cell.** This command (which can also be activated by pressing ENTER rather then **E** or **e**) causes a cursor to appear in the highlighted cell and allows the value to be edited. Press ENTER to register the edited value and return to the menu. Alternatively, if there is at least one undefined value in the data (shown as ?), press TAB to register the edited value and move immediately to the next undefined value. A blank value cannot be entered, but the value does not have to be a number even if it is entered in a cell where a number would normally be expected.

5. **New v column.** All entries in the v column will be replaced by undefined values (shown as ?). This is to facilitate entry of a new set

of data with the same or nearly the same experimental design (same set of independent variable values) as an existing set. (If the dependent variable is not called v the actual name will appear in the menu.)

6. **Restore v column.** After selection of the preceding entry, this causes all undefined values in the v column to be restored to the values they had before **N** was pressed. Values that were entered after **N** was pressed will not be restored to their previous values.

7. **Insert weights column.** If a column of weights does not already exist, this command causes a column of weights to be added, with the name w and a value of 1.0000 in each line. The values can then be edited at will.

8. **Kill weights column.** If a column of weights exists it can be deleted with this command.

9. **Weighting function.** This command allows a function for calculating weights to be defined (Section 9.2.2).

10. **Set error parameters.** If parameter values have been calculated, this option allows the error function used in the next entry to be specified. Errors are calculated with variances given by the equation

$$\sigma^2 = \sigma_0^2 + \sigma_2^2 \, \hat{v}^2. \tag{9.1}$$

Consequently, it is necessary to enter values for σ_0 and σ_2, the two standard deviations, and for the outlier percentage and outlier factor. Outliers will be generated with the specified frequency and with standard deviations given by the above equation but multiplied by the specified factor. (If this option has not been chosen explicitly, it is activated automatically at the first selection of **Simulate**).

11. **Simulate.** If parameter values have been calculated, this option causes the values of the dependent variable to be replaced by values calculated with the fitted equation with random errors calculated as described in the preceding entry.

12. **Exit editor.** Return to the DATA MENU. Note that in this menu ESCAPE is the same as **Exit**: it does not restore the data to the state before entry to the menu. If you need to do this you can read the original file again (provided it has been saved).

13. TAB moves the highlighting to the next undefined value (shown as ?), if there is one.

9.2.2   Weighting function window

This window opens when the command Weighting function is selected in the EDITING MENU (Section 9.2.1). Its purpose is to allow you to define a function for calculating weights according to the principles advocated by Mannervik, Jakobson, and Warholm (1986). For variables [S] and v, the function has the following form:

$$w = a + \frac{b + c[\mathrm{S}]^d + ev^j}{g + h[\mathrm{S}]^i + jv^k}, \tag{9.2}$$

where the parameters $a, b, \ldots k$ are all adjustable. The coefficient $a, b, c, e, g, h,$ and i can have any non-negative values; the exponents can have values in the range 0.25 to 4; unacceptable values are ignored. In addition, the denominator as a whole must not be zero. Exponents can be defined for terms with zero coefficients, but if they are they have no effect on the calculation and a ? appears after such a value in the display. The upper part of the window shows an interpretation of the current values: the more obvious redundancies are recognized (for example if the denominator is a constant it will not be shown but the numerator coefficients will be scaled appropriately), but less obvious ones (such as defining a numerator that is a constant multiple of the denominator) are not.

Move between cells using the arrow keys ↑ ↓, TAB (one column to the right), and SHIFT-TAB (one column to the left). Press ENTER only to enter the complete function, i.e. only when all the individual values are set. The individual weights can then be edited manually if you wish—although they may be created with a definite function they are not permanently linked to that function.

9.2.3 Data type window

This window opens automatically when you finish entering a new set of data, and it can also be called with the command **R**edefine data type in the DATA MENU. Its function is to define the kind of data entered in each column of the data file, so that Leonora knows what kinds of equation can be fitted. The left-hand panel lists the possible data types and the keys to press to select them. The right-hand window initially contains the name of the variable in column 1. After a data type has been selected it is displayed, and the next variable is shown in row 2, and so on until all variables have been defined. At this point you can press ENTER to accept the definitions shown,* or any other key to start again. If the definitions are not acceptable (for example if no dependent variable is specified, or you try to specify more than one dependent variable) an error message will appear.

---

*The definitions are only used at the moment of selecting an equation. A consequence of this is that if you fit an equation to some data, then redefine the data type after selecting **R**edefine data type, and then return to do some calculations without selecting a new equation, the original equation and data types will still be in force.

The possible definitions are as follows.

1. **Substrate concentration.** Any variable x for which the expression for the dependent variable is of the form $ax/(b + cx)$, where a, b, and c are constants or functions of other independent variables.

2. **Inhibitor concentration.** Any variable x for which the expression for the dependent variable is of the form $a/(b + cx)$, where a, b, and c are constants or functions of other independent variables.

3. **Activator concentration.** Any variable x for which the expression for the dependent variable is of the form $ax/(b + cx)$, where a, b, and c are constants or functions of other independent variables.*

4. **pH.** Any variable x for which 10^x behaves like an inhibitor or an activator (or both together).

5. **Linear independent.** Any variable x for which the expression for the dependent variable is of the form $a + bx$, where a and b are constants or functions of other independent variables.†

6a. **Dependent variable.** The variable on the left-hand side of the equation.

6b. **Dependent variable.** The variable on the left-hand side of the equation (alternative key allowed as the dependent variable is usually represented by the symbol v in steady-state kinetics).

7. **Time as dep. variable.** This is a special type that can be used *only* for fitting the integrated Michaelis–Menten equation, i.e. it is for defining a variable as t in the equation $V^{app.}t = s_0 - s + K_m^{app.}\ln(s_0/s)$ when some other variable is defined as s.

9.3   Equation menu

This menu is activated by selecting **Equation** in the MAIN MENU. It differs from other menus in containing some entries that vary according to the kind of data and others that are the same for all kinds of data. The latter are described first.

9.3.1   Common entries in the equation menu

1a. **Show equations algebraically.** Equations are normally listed by name, but this command causes the actual equations to be displayed. The

---

*Although this is algebraically the same as the definition of a **Substrate** concentration, Leonora does not allow them to be used interchangeably.

†You cannot mix linear and hyperbolic independent variables in the same equation.

symbols for variables are those you have defined in the data. Thus although in Section 9.3.3 below the symbols [S] and v are used for substrate concentration and rate respectively, they will appear as Conc and rate in the equation if these are the symbols used when entering the data.

1b. **Show equations by name.** This is the converse of the previous entry and appears if it is de-activated (and vice versa). It restores the normal display.

2. **New equation.** This allows a new equation to be defined and, if appropriate, added to the data file so that it can be recalled in the future (Section 9.3.3).

3. **Debugging mode.** This command has no immediate effect, but has the delayed effect of causing the matrices corresponding to the selected equation to be displayed after the equation is selected (Section 9.3.3 explains these matrices). It is also necessary to set debugging mode in order to delete an equation from the menu (next paragraph).

4. *Deleting an equation.* There is no specific menu command, and the procedure has deliberately been made a little cumbersome as a guard against accidental deletions. To delete an equation, first set debugging mode, then select the equation you want to delete, and press the DELETE key when the corresponding matrices appear. When the warning appears, press DELETE again to confirm the deletion. To change your mind, do not press DELETE a second time but press any other key. Deleting an equation does not, in fact, delete the information about it from the file of equations, but just inserts XXXX at the beginning of the line containing the information about the type of data required. Thus although you cannot reinstate a deleted equation from within Leonora, it is easy to do so by editing the file PRIVATE\EQUATION.DAT (Section 10.6).

5. **Exit.** Return to the MAIN MENU.

9.3.2   Specific entries in the equations menu

These appear at the beginning of the menu, i.e. before the entries listed in the previous section. The triangle ▶ is initially placed by the last equation, i.e. if you press ENTER you will normally select the most complex equation available. However, if you make no selection at all (either because you don't call the EQUATIONS MENU at all, or because you exit without making a selection), you will select the equation most commonly used in the literature for the particular data type, as indicated by the parentheses (*default*) in the next sections.

Table 9.1 Algebraic forms of the menu entries of the equations listed by name in Section 9.3.2

"Show equations by name"	"Show equations algebraically"	
Michaelis–Menten	**M:**	$v = V \cdot [S] / (Km + [S])$
Substrate inhibition	**S:**	$v = V \cdot [S] / (Ks + [S] (1 + [S] / Ksi))$
Michaelis–Menten (ignoring [I])	**M:**	$v = V \cdot [S] / (Km + [S])$ ignoring [I]
Primary Michaelis–Menten (at each [I])	**P:**	$v = V(app) \cdot [S] / (Km(app) + [S])$ at each [I]
Generic inhibition (at each [S])	**G:**	$v = v^\circ / (1 + [I] / Ki)$ at each [S]
Competitive inhibition	**C:**	$v = V \cdot [S] / (Km (1 + [I] / Kic) + [S])$
Uncompetitive inhibition	**U:**	$v = V \cdot [S] / (Km + [S] (1 + [I] / Kiu))$
Mixed inhibition	**I:**	$v = V \cdot [S] / (Km (1 + [I] / Kic) + [S] (1 + [I] / Kiu))$
Michaelis–Menten (ignoring [B])	**M:**	$v = V \cdot [A] / (Km + [A])$ ignoring [B]
Michaelis–Menten (ignoring [A])	**I:**	$v = V \cdot [B] / (Km + [B])$ ignoring [A]
Primary Michaelis–Menten (at each [B])	**P:**	$v = V(app) \cdot [A] / (Km(app) + [A])$ at each [B]
Primary Michaelis–Menten (at each [A])	**R:**	$v = V(app) \cdot [B] / (Km(app) + [B])$ at each [A]
Substituted-enzyme mechanism	**S:**	$v = V \cdot [A] \cdot [B] / (KmB \cdot [A] + KmA \cdot [B] + [A] \cdot [B])$
Ternary-complex mechanism	**T:**	$v = V \cdot [A] \cdot [B] / (KAB + KmB \cdot [A] + KmA \cdot [B] + [A] \cdot [B])$
Ordered equilibrium mechanism	**O:**	$v = V \cdot [A] \cdot [B] / (KAB + KmB \cdot [A] + [A] \cdot [B])$
S-shaped pH profile	**S:**	$k = klim / (1 + [H+] / K1)$
Z-shaped pH profile	**Z:**	$k = klim / (1 + K2 / [H+])$
Bell-shaped pH profile	**B:**	$k = klim / (1 + [H+] / K1 + K2 / [H+])$

Each entry is shown only by name in paragraphs 1–4 below, but all of the corresponding algebraic forms are listed in Table 9.1. What you will actually see on the screen will depend on whether you have selected Show equations algebraically (Section 9.3.1) or not.

1. *Equations available for concentration–rate data*

   (a) Michaelis–Menten (*default*).

   (b) Substrate inhibition.

2. *Equations available for inhibition data*

   (a) Michaelis–Menten (ignoring [I]). Use this option if you want to treat the data as ([S], v) data, i.e. behave as if the column of [I] values did not exist. In practice, this is appropriate if the effect of the inhibition is so weak that you are not sure whether it is significant.

   (b) Primary Michaelis–Menten (at each [I]). Use this option if you want to treat the data as a series of experiments measuring the dependence of v on [S] at each [I] value.

   (c) Generic inhibition (at each [S]).

   (d) Competitive inhibition (*default*).

   (e) Uncompetitive inhibition.

   (f) Mixed inhibition.

3. *Equations available for two-substrate data*
   (In this paragraph, and in the corresponding part of Table 9.1, the two substrate concentrations are assumed to be [A] and [B].)

   (a) Michaelis–Menten (ignoring [B]).

   (b) Michaelis–Menten (ignoring [A]).

   (c) Primary Michaelis–Menten (at each [B]). Use this option if you want to treat the data as a series of experiments measuring the dependence of v on [A] at each [B] value.

   (d) Primary Michaelis–Menten (at each [A]).

   (e) Substituted-enzyme mechanism. This is the type of equation sometimes called ping-pong.

   (f) Ternary-complex mechanism (*default*).

   (g) Ordered equilibrium mechanism. This equation is not normally available when Leonora is installed, but is listed here because it is used to illustrate how to add a new equation to the menu (Section 9.3.3).

4. *Equations available for pH data*

   (In this paragraph, and the corresponding part of Table 9.1, the variables are assumed to be pH and k, which can be any pH-dependent parameter, with a value k_{lim} for the fully activated state. Although you input pH values directly, and these are displayed after the results are calculated, the actual variable used in the calculations is 10^{-pH}, which is displayed in the equations as [H+]).

   (a) S-shaped pH profile. Use this if k increases as the pH increases.

   (b) Z-shaped pH profile. Use this if k decreases as the pH increases. The symbol K2 is used for the parameter (Table 9.1) to preserve the correspondence with the next entry.

   (c) Bell-shaped pH profile (*default*).

5. *Equation available for concentration–time data.* Only one equation is provided, the integrated Michaelis–Menten equation $V^{app}{\cdot}t = a_0 - a + K_m^{app}{\cdot}\ln(a_0/a)$, and others cannot be added (because this equation is treated as a special case in the calculations).

6. *Equations available for linear x–y data.* If x (or t, in line (d)) is an independent variable for a linear equation and y is the dependent variable, the following equations are available.

   (a) Straight line, $y = b_0 + b_1 x$

   (b) 2: Quadratic, $y = b_0 + b_1 x + b_2 x^2$

   (c) 3: Cubic, $y = b_0 + b_1 x + b_2 x^2 + b_3 x^3$

   (d) First-order (*default*), $\ln y = \ln y_0 + kt$.

7. *Equations available for linear x–x–y data.* If x_1 and x_2 are linear independent variables and y is the dependent variable, the following equations are available.

   (a) Only 1st variable, $y = b_0 + b_1 x_1$

   (b) Only 2nd variable, $y = b_0 + b_2 x_2$

   (c) All variables (*default*), $y = b_0 + b_1 x_1 + b_2 x_2$.

8. *Equations available for linear x–x–x–y data.* If x_1, x_2, and x_3 are linear independent variables and y is the dependent variable, the following equations are available.

   (a) Only 1st variable, $y = b_0 + b_1 x_1$

   (b) Only 2nd variable, $y = b_0 + b_2 x_2$

   (c) Only 3rd variable, $y = b_0 + b_3 x_3$

(d) **4**: Variables 1 and 2, $y = b_0 + b_1 x_1 + b_2 x_2$

(e) **5**: Variables 1 and 3, $y = b_0 + b_1 x_1 + b_3 x_3$

(f) **6**: Variables 2 and 3, $y = b_0 + b_2 x_2 + b_3 x_3$

(g) All variables, $y = b_0 + b_1 x_1 + b_2 x_2 + b_3 x_3$.

9.3.3 Adding a new entry to the equations menu

This section illustrates how to use the command New equation in the
EQUATIONS MENU, taking as an example the ternary-complex mechanism
equation (Section 9.3.2, paragraph 3(f)) with the term in [B] missing from
the denominator. This is the equation that applies if A and B bind in that
order and both binding steps are assumed to be at equilibrium in the
steady state.

When New equation is selected a window entitled ENTER NEW EQUATION
appears, with the substitutions «1» [A] «2» [B] «3»v shown at the bottom.
You do not need to type these substitutions, but they are used for storing
the equation for future use in a generic form that does not depend on the
particular names you are currently using for the variables, i.e. the equation
will still be available if in a future data set you choose to replace [A] and
[B] by [NAD] and [Glc] respectively. Multiplication signs must be
included explicitly as *, and parentheses must be correctly balanced
(though a missing final parenthesis may sometimes be acceptable).

To define the equation

$$v = V [A] [B] / (KAB + KmB [A] + [A] [B])$$

you should type

$$v = V * [A] * [B] / (KAB + KmB * [A] + [A] * [B])$$

Note that v = is already displayed when you begin. Although you can edit
this, you are strongly advised not to do so because you will almost
certainly produce an equation that will not be accepted. The equation can
contain up to 255 characters, of which only 48 appear on the screen at the
same time. It is best not to make the equation longer than necessary,
however, because unnecessarily long parameter names will usually need to
be truncated in displaying results. One should also avoid approaching the
limit of 255 characters too closely, because the stored equation will
contain spaces before and after + and = signs even if you do not type
them, and consequently may be longer than the equation as you type it.

When you press ENTER, the new equation will appear twice more on the screen:

$$«3»=V * «1» * «2»/ (KAB + KmB * «1»+ «1» * «2»)$$

$$v = V \cdot [A] \cdot [B] / (KAB + KmB \cdot [A] + [A] \cdot [B])$$

Although the third line should be exactly the same was what you typed on the first line, apart from possible insertion and deletion of spaces, and substitution of $*$ by $\cdot$, you should check it all the same because it represents the equation after it has been interpreted by Leonora, i.e. the equation as typed in has first been converted to a set of matrices suitable for storing and calculating, then converted back into a form suitable for display; it is not just the same string with some minor substitutions. More important (because errors are more likely to be evident there), you should check that all of the variable names have been substituted according to the substitutions listed at the bottom of the window. If any have been left unsubstituted, they will be treated as parameter names. Such cases are most likely to result from failure to type the names in exactly the same form as they have been defined. For example, if you have typed [a] instead of [A] it will be interpreted as a parameter symbol.

After checking that the equation is correctly typed, press ENTER, or any other key if you want to edit the equation before entering it. A window will appear showing the matrices corresponding to the new equation. The title of the window contains the interpreted equation (as on line 3 of the previous window) preceded by the word Hyperbolic if Leonora interprets the equation as a generalized Michaelis–Menten equation, or Linear if it interprets it as a linear equation. The matrix entries allow a further check that the equation has been interpreted correctly; they refer to the *reciprocal* form of the rate equation, which in this case is

$$1/v = (KAB/V) * (1/[A]) * (1/[B]) + (KmB/V) * (1/[B]) + (1/V)$$

There are three terms on the right-hand side of this equation, and thus three lines in each matrix. In the column V the -1 in each line indicates that $1/V$ is a factor in each parameter of the reciprocal equation; in the column KAB, the 1 in line 1 indicates that KAB is a factor in the first parameter 0 in lines 2 and 3 indicates that it does not appear in parameters 2 and 3, and so on. In a similar way the right-hand matrix shows that $1/[A]$ is a factor in the first variable, and that $1/[B]$ is a factor in variables 1 and 2. If Leonora detects an error at this stage it will show an error message in the right-hand panel: for example Error:2 for term 3 would indicate an equation in which the combination of independent variables in the third term was exactly the same as for the second. This is not allowed because such equations lead to indeterminate solutions.

When you press any key to remove the matrix window, the query Add it to the menu (Y/N)? will appear. If you answer **no**, the equation will be available for use with the current set of data, but will not be stored for future use. If you answer **yes**, the line Enter model name: will appear, after which you can type *Ordered equilibrium mechanism* ENTER. Then in response to the message Press key to be high-lighted in the menu press **o**. Note that you cannot choose a key such as **t** that is already needed for another entry in the menu, as indicated by the warning Don't choose one in use already (DIMNPQRSTUX).

9.4 Output requirements menu

1. **No output file** (*default*). With this option no file of results is written. Do not set this option if you intend to send the results of the calculation to the printer (other than by using the PRINT-SCREEN key): see Section 9.5, paragraph 8.

2. **Results.** For a data set called DATAFILE.MMD, this option causes a file DATAFILE.OUT to be written during execution, which contains all of the results that are sent to the screen during the calculation, in essentially the same format. You should set this option or the one following if you intend to send the results of the calculation to the printer (Section 9.5).

3. **Details.** For a data set called DATAFILE.MMD, this option generates a file DATAFILE.OUT, which contains all of the information produced by select-ing **Results**, and, in addition, many details of intermediate stages in the calculations. This option is intended to allow a calculation to be followed in detail, either as a way of checking for errors or for better understanding how the final results are arrived at.

4. **Exit.** Return to the MAIN MENU.

9.5 Calculations menu

1a. **Calculate best fit.** Calculate the parameter values for the equation that has been selected. Examples of the results screens that appear are given in Chapter 8.

1b. **Calculate best fit for [B] = 1.0000.** The previous entry appears in this form if a primary-plot equation has been selected (see Section 9.3.2, paragraphs 2(b)–(c) and 3(c)–(d)); the example assumes that the 'constant' variable is [B] and that one of its values is 1.

2. **Next [B].** This entry appears only if a primary-plot equation has been selected (see Section 9.3.2, paragraph 2(b)–(c) and 3(c)–(d)), and the

example assumes that the 'constant' variable is [B]. Selecting it causes the constant shown in the Calculate best fit for [B]= ... entry to be changed to the next available constant value. If the last one has been reached it cycles back to the first.

3. Define method. Choose the method of fitting to be used (Section 9.5.1).

4. Define weighting system. Choose the system of weighting to be used with either the least-squares or least-absolutes methods (Section 9.5.2).

5. Display speed. Select between the following commands, which are also available while a calculation is proceeding, i.e. N, F, S, or W can be pressed after selecting Calculate best fit in order to alter the speed of calculation.

   (a) No display (no intermediate results appear on the screen).

   (b) As fast as possible (display intermediate results, but no pause between them).

   (c) Slow enough to see (pause of 0.5 s between steps).

   (d) Wait for key (pause after each step until a key is pressed).

6. Hill exponents. This causes a window to appear listing each of the independent variables in the model, with a value defined initially as 1.0000. Each of these values can be edited to a value in the range 0.25 to 4, and in the subsequent calculations the variable will be raised to the power entered.* For example, if an entry [S]:1.0000 is changed to [S]:2.500, each occurrence of [S] in the equation will be treated as if it were $[S]^{2.5}$.

7. Jackknife. Once a calculation has been done with a data set of n observations, repeat it n times with each of the n data sets of $(n-1)$ produced by omitting one observation each time, and estimate parameter values from the $(n+1)$ sets of results. Remember that if one calculation takes a significant amount of time a jackknife calculation will take a lot of time!

8. Print results. Provided that an output file exists (Section 9.4), and provided that a printer is connected, the results of the last calculation will be printed. Leonora will accept the command even if no printer is connected, because some computers take a very long time (over a minute) to recognize that a printer is not connected, and on such computers checking for the existence of a printer every time the

---

*Leonora does not optimize Hill exponents; it only allows values to be entered as constants.

program is run would waste a lot of time. If you give the command when there is no printer, you will lose as much time as your computer needs before it is sure there is no printer.

9. Show parameters. Return to the list of parameter values already displayed after a calculation.

10. Show statistics. For least squares with robust weighting this will generate a window containing the values described in Section 8.8. To make it clear what has actually been calculated, all of these values are labelled with algebraic expressions rather than names. If robust weights are not used the lines referring to robust weights w_2 are omitted. For least absolutes or median calculations, sums of absolute values are shown instead of sums of squares. If the data include some replicate observations, the window will also show the results of calculation of sums of squares for pure error (determined from the replicates) and lack of fit (residual sum of squares corrected for pure error). If this selection is made after using dynamic weights, the statistical data will appear after a warning `Cannot compare results between models with dynamic weights`, and all of the lines containing sums of squares will be displayed in square brackets.

11. Show results. Return to the list of results already displayed after a calculation.

12. Save method/weights. Save the currently selected method and weighting system for use by default in the future.

13. Exit. Return to the MAIN MENU.

9.5.1 Methods menu

1. Least squares (*default*). Obtain the parameter values that minimize a function of the form $SS = \sum W(v - \hat{v})^2$, where W is a weight defined using the WEIGHTING SYSTEM MENU (Section 9.5.2).

2. Least absolutes. Obtain the parameter values that minimize a function of the form $SA = \sum w^{1/2}|v - \hat{v}|$, where w is a weight defined using the WEIGHTING SYSTEM MENU (Section 9.5.2).

3. Median. Calculate all possible sets of p parameter values given by each combination of p observations in turn, and set each parameter to the median of the values obtained. Note that this implies a lot of calculation if p is greater than 2 and n is large, for example for $p = 4$ parameters and $n = 36$ observations there are $58\,905$ combinations to be tried, for $p = 4$ and $n = 100$ there are $3\,921\,225$ combinations, and so on. However, once started a calculation can be interrupted by pressing ESCAPE. As the combinations are sampled in a pseudo-random

order even a small sample of the total set may still yield reasonably representative results. A calculation cannot be started if n^p is greater than 2^{30} (about 10^9). If a calculation terminates normally, or is interrupted by pressing ESCAPE, all temporary files created during the calculation are deleted. However, if Leonora crashes during it or you break in by pressing CTRL-BREAK, files TEMPFILE.1, TEMPFILE.2, and so on may be left behind.

4. Graphical. This option is available only if you return to the CALCULA-TIONS MENU after plotting the results of a previous calculation. In this case you can use it to obtain the calculated values of the dependent variable that correspond to the parameter values at exit from the PLOTTING MENU (Section 9.6).

5. Minimax. This method sets out to minimize not the average deviation, as with Least squares or Least absolutes, but the largest individual deviation. This is *not* a statistical method, and is not appropriate if the discrepancies between observed and calculated values are statistical in origin (i.e. if they result from random experimental error). It is appropriate if you want to obtain a simple approximation to a function that can be calculated exactly but which is tedious to calculate exactly. This method requires a set of starting guesses obtained by some other method.

6. External. This option allows the parameter values (e.g. values obtained with another program) to be entered from the keyboard.

7. Exit. Return to the CALCULATIONS MENU.

9.5.2 Weighting system menu

In this menu the sign $\sqrt{}$ appears opposite whichever (if any) of the first four options is in force.

1. Dynamic weights (*default* when the method is least squares, unless external weights exist). Assess the appropriate weighting system from internal evidence in the data.

2. Fix current weights. Convert the dynamic weights currently in use into constant weights. This option must be selected before meaningful statistical comparisons are possible between equations fitted with dynamic weights.*

3. Uniform weights for v. Assign a weight of 1 to each v value and keep it constant. This assumes that each v has the same standard deviation.

---

*This command adds information to the data file, and in consequence an asterisk * is displayed on the top line of the screen (Section 8.9) after return to the MAIN MENU.

4. **Intermediate weights.** Calculate weights intermediate between uniform and relative. Each weight is set to $1/\sigma^2$, the variance of v, which may be calculated from either of the following two equations:

$$\sigma^2 = \sigma_0^2 + \sigma_2^2 \hat{v}^2 \ (default), \text{ or } \sigma^2 = K\hat{v}^\alpha,$$

in which σ_0 and σ_2 or K and α are constants. In the first case, only the ratio σ_0/σ_2 is actually needed for calculating weights; this has the same dimensions as v and is referred to simply as σ/σ on the screen. In the second case only α (not K) is needed for calculating weights. Values of $\sigma_0/\sigma_2 = \infty$ and $\sigma_0/\sigma_2 = 0$ correspond exactly to $\alpha = 0$ and $\alpha = 2$ respectively, and also to **Uniform** weights for v and **Relative** weights for v respectively. Intermediate values of σ_0/σ_2 and α do not correspond exactly to one another, but an approximate correspondence is given by the following equation, which is used by Leonora for interpolating between values of σ_0/σ_2, because direct interpolation on a scale that goes from 0 to ∞ is not very convenient:

$$\alpha \approx \frac{2}{1 + (\sigma_0^2/\sigma_2^2)/\bar{v}^2},$$

where $\bar{v}$ is the mean value of v.

The INTERMEDIATE WEIGHTS MENU contains the following entries.

(a) **Nearly uniform.** This defines $\alpha = 0.5$ or the value of σ_0/σ_2 that gives $\alpha = 0.5$ in the approximate equation above.

(b) **Middle of range.** This defines $\alpha = 1$ or the value of σ_0/σ_2 that gives $\alpha = 1$ in the approximate equation above.

(c) **Nearly relative.** This defines $\alpha = 1.5$ or the value of σ_0/σ_2 that gives $\alpha = 1.5$ in the approximate equation above.

(d) **Enter value of σ/σ** (or **Enter value of α**). This allows a particular value of the weighting constant to be entered.

5. **Relative weights for v.** Assign a weight of $1/\hat{v}^2$ to each v value (where $\hat{v}$ is the calculated value of v). This assumes that each v has the same coefficient of variation.

6a. **A: Use α weights.** Calculate intermediate weights in terms of the parameter α. As the opposite (σ_0/σ_2) weights are in force when this option is available the symbol σ appears opposite the entry.

6b. **S: Use σ/σ weights** (*default*). Calculate intermediate weights in terms of the parameter σ_0/σ_2. As the opposite (α) weights are in force when this option is available the symbol α appears opposite the entry.

7a. Use observed v. Calculate weights with v in place of $\hat{v}$ in all expressions. In many cases this will lead to faster calculation than using $\hat{v}$, but it is not theoretically justifiable. As the opposite Use calculated v convention is in force when this option is available the symbol c appears opposite the menu entry.

7b. Use calculated v (*default*). Calculate weights with $\hat{v}$ where appropriate. As the opposite Use observed v convention is in force when this option is available the symbol o appears opposite the menu entry.

8. Reset biweight parameter. This calls the ROBUST WEIGHTING MENU, which allows the parameter c (normally 6 or ∞) used in the biweight calculation to be reset. The value currently in force is shown at the bottom of the menu window. A value of ∞ is equivalent to not using the biweight adjustment, i.e. to ordinary non-robust least squares. Robust weighting causes the ordinary weights w, defined either in the data or with the parameters σ_0/σ_2 or α, to be replaced by robust weights W using the following formula:

$$W = \begin{cases} w(1 - u^2)^2 & \text{if } |u| < 1 \\ 0 & \text{if } |u| \geq 1 \end{cases},$$

in which u is a measure of the deviation of the observed value from the best-fit model:

$$u = w^{1/2}(v - \hat{v})/cS,$$

in which S is the median absolute value of $w^{1/2}(v - \hat{v})$ and c is the constant whose value is set by this menu.

If the value of c is suitably chosen, this transformation of the weights has almost no effect on the weighting of observations showing small to moderate deviations from the calculated values, but substantially decreases the weighting of highly deviant observations (in limiting cases to zero). The menu provides the following possibilities.

(a) Standard (*default*). This is the recommended value, $c = 6$.

(b) Tolerant. This sets $c = 12$, and has the effect of decreasing the robustness, i.e. making the weighting more like ordinary least squares.

(c) Intolerant. This sets $c = 3$, and makes the weighting very intolerant of deviant observations. Values any smaller than this are likely to produce unpredictable and arbitrary results, because the observations that Leonora considers highly deviant will then be those that happen to give large deviations in the first cycle of calculation, when the weights w are unlikely to be well calculated.

(d) Enter c. This allows the value of c to be entered directly.

(e) Don't use robust weights. This is equivalent to making c infinite, and means that the robust transformation is omitted. In other words, the method becomes ordinary least squares, the only possibility available in most programs.

(f) Exit. Return to CALCULATIONS MENU.

9. External weights (*default* when external weights exist, not available otherwise). If a weights column exists in the data, use it to assign weights. Leonora assumes that if you have defined weights then you want to use them, and therefore takes this option as default if a weights column exists.

10. Lineweaver–Burk. Use a weight of 1 for each value of $1/v$. Why anyone would wish to throw away most of the information in the data by using weights that are likely to be grossly far from the correct weights is a mystery to me, but Leonora has been written in the belief that users of programs should be allowed to do what they want to do, not just what the programmer thinks they ought to want. However, this belief does not extend to the point of making it easy to make inappropriate choices, and so this option cannot be saved as default but must be explicitly selected each time.

11. Woolf. Use a weight of 1 for each value of a/v. See the note for the preceding entry, which also applies here, albeit with less force.

12. Exit. Return to the CALCULATIONS MENU.

9.6 Plotting menu

1. Axes. Select the axes to be used. Use the TAB key to toggle between Abscissa and Ordinate. In each case, select a variable with ← → and vary the power to which it is to be raised (-1, 0, or 1 only) with ↓ ↑. If the axes are set so that the calculated values fall on straight lines, the message LINEAR is displayed; if a plot in parameter space consists of straight lines, the message DIRECT PLOT POSSIBLE is displayed. If you select an axis for the abscissa that contains more than one independent variable, a window will appear asking which variable is to be used for the calculations.

2. Fit by eye. Alter the current parameter values by small arbitrary amounts. This is intended for fitting the data by eye without being biassed by knowledge of the calculated values.

3. Scale ranges. Select the ranges to be used for plotting. Use ↓ ↑ to toggle between Abscissa and Ordinate, and TAB to toggle between

left (lower limit) and right (upper limit) columns. Pressing **z** is a quick way of extending the range so that it includes zero, moving the cursor immediately to the other axis; for example, if the cursor is in the Abscissa line and the range displayed is from 0.100 to 10.900, pressing **z** changes the range to from 0.000 to 10.900 and moves the cursor to the Ordinate line.

4. Calculation range. Select the range of the independent variable to be used for the calculation (if this step is omitted the range defined for the abscissa scale will be used).

5. Plot. Draw the plot for the axes and scales selected. When you next press a key the GRAPHICAL MENU (Section 9.7) will appear. Press **X** to escape from this and return to the PLOTTING MENU.

6. Residual plot. Draw a residual plot, keeping the abscissa axis as it has been defined, but replacing the ordinate values by the differences between observed and calculated values of the specified ordinate. Thus, if the ordinate axis has been defined as a/v, the residual plot will actually plot values of $(a/v) - (a/\hat{v})$.

7. Save axes as default. Save the currently defined axes so that they will be used as the default in future. *Note* that each data type has its own default; thus, changing the default axes for concentration–rate data has no effect on the default axes for inhibition data.

8. Direct linear plot. If axes have been selected such that the loci of parameter values that satisfy the observations exactly are straight lines the message DIRECT PLOT POSSIBLE will have been displayed in the **Axes** menu, and this option causes such a plot to be made.

9. No menu space/Make menu space. Plots are normally made leaving enough space to display the graphical menu without overlapping the plot. However, selecting No menu space makes the whole height of the screen available for the plot (at the expense of having a menu that is difficult or impossible to read). Make menu space restores the default.

10. Exit. Return to the MAIN MENU.

9.7   Graphical menu

This menu appears on the screen at the same time as any plot (unless you have initially suppressed it using the option No menu space in the PLOTTING MENU, Section 9.6). Because the computer is no longer in text mode its appearance is different from those of the other menus in Leonora. At the left it lists the current parameter values and the gain, i.e. the amount by which parameter values are changed each time ↑ or ↓ is pressed; at the right it contains the following entries.

1. Tab: Select parameter (*). The TAB key cycles through the parameters to determine which is affected by the arrow keys. The (*) is a reminder that the selected parameter is marked by * in the display at the left (on a colour screen it is also shown in a different colour).

2. Arrows: Adjust selected parameter. Alter the value of the selected parameter by multiplying ($\uparrow$) or dividing ($\downarrow$) by the gain. After a value is changed the old value is erased from the screen but the new value is not displayed unless 2 s elapses without a key being pressed. (This is to allow rapid redrawing of the lines, as displaying the value would produce a perceptible delay.)

3. PageUp: Increase gain. If the gain is less than 2, increase it to its square. The gain is initially 1.001 for residual plots, 1.01 otherwise: the maximum is thus about 2.78 for residual plots, 3.57 otherwise. The old value of the gain is erased from the screen, but the new value is not displayed until a new parameter is selected by pressing TAB.

4. PageDown: Decrease gain. Decrease the gain to its square root. The old value of the gain is erased from the screen, but the new value is not displayed until a new parameter is selected by pressing TAB.

5. H: Hard copy. Send numerical values of the coordinates of points and lines to a file DATAFILE.PLT (if DATAFILE.MMD is the name of the data file).

6. X: Exit. Return to the PLOTTING MENU.

9.8 Setting defaults

Calling the DEFAULTS MENU is an invisible option in the MAIN MENU—invisible to avoid complicating this menu with an option that it is not appropriate to call very often. It is done by pressing * at the MAIN MENU, and results in a menu with the following entries.

1. Colours. Define the colours (or intensities, on a monochrome screen) to be used for presenting different kinds of information. The adjustments are made as follows.

   (a) Using the PAGE-UP and PAGE-DOWN keys, move the word ADJUST (near the bottom-left corner of the screen) level with Normal text, Highlighted, Bright background, Faint, Warning, or Help screen. Then use $\leftarrow$ $\rightarrow$ to adjust the foreground colour for the selected context, and $\downarrow$ $\uparrow$ to adjust the background colour. Examples of the different contexts appear in the top half of the screen. An example of Results also appears, but this cannot be adjusted independently: the background is the same as that of Help screen; the foreground is the same as that of Warning.

(b) The bottom-right corner of the screen shows all possible combinations of background and foreground. (On a VGA or EGA screen there should be 8×16 combinations, all different; on a monochrome or CGA screen not all combinations will be different.) When satisfied, press ENTER (or ESCAPE to restore the defaults in force at entry to the menu).

(c) Although Normal text, Highlighted, and Faint can be adjusted independently, it is best to choose the same background for all three (preferably black, dark blue, or grey unless you like a very gaudy display), and the menus will look quite peculiar if you do not.

2. Cursor. Set the values of the scan lines for the cursor. Unfortunately, different 'compatible' computers differ much more than they should in how they interpret the values that can be defined within a program. In consequence you may find a lot of flickering at times when no cursor should be visible (e.g. in a menu) or the cursor may be invisible when it ought to be visible. Call this adjustment to correct either of these conditions.

(a) The first cursor to appear should look like the cursor at the DOS prompt. If it does not, adjust it with ↓ ↑ for the bottom scan line and PAGE-UP and PAGE-DOWN for the top scan line. The numerical values corresponding to the scan settings will be shown, but the relationship between these and the appearance of the cursor may well be obscure; at least, it is to me.

(b) Pressing TAB changes to a cursor that should look thicker than normal; if it does not, adjust it in the same way.

(c) Pressing TAB again changes to an invisible cursor. If it is not invisible you will see flickering not only in the cursor window but also at the top of the screen. Adjust in the same way as the others; then press ENTER to return to the DEFAULT MENU.

3. Output file (NONE/SHORT/FULL). This brings up the OUTPUT MENU (Section 9.4) and defines the default to be used when that menu is *not* called from the MAIN MENU. The parentheses indicate the current setting (not the current default).

4. Beep (ON)/Beep (OFF). Disable or enable the beep that Leonora makes when giving a warning.

5. Time out (180s). Set the time that must elapse before Leonora enters screen-saving mode. The parentheses show the current setting. In screen-saving mode pressing any key restores the normal display, *except* when a graph is on the screen, in which case pressing a key returns to the PLOTTING MENU. Entering a time of 0 disables the

function altogether: this may be advisable if you have another screen-saving program resident in memory.

6. Automatic help (ON)/Automatic help (OFF). Determine whether the bottom line of the screen should be used for help information. Switching to ON has an immediate effect; switching to OFF does not suppress the automatic help until exit from the DEFAULT MENU.

7. Editor. Define the behaviour of the editor. When editing values in Leonora you may prefer to adjust the editor to behave like other editors you use. In all cases select the line containing the feature you want to change with ↑ ↓ and toggle the setting with ← →. The following possibilities exist.

   (a) Automatic conversion of lower- to upper-case. In some contexts (e.g. entering a DOS file name) this option will be taken as YES regardless of what you enter here, but in most contexts it can be set as you wish.

   (b) Cancel input value if non-arrow key typed first. If this is set to YES the editor will assume that you are entering a new value rather than editing an existing one unless you press an arrow key before typing anything else.

   (c) Start editing with insertion mode. While in the editor you can use the INSERT key to toggle between *insertion mode* (anything typed is added to what is already there) and *overwrite mode* (anything typed overwrites whatever is already there). This option decides which mode you will start in if you do not press INSERT.

   (d) Start with cursor at the END of the value. This determines where the cursor is when you start editing an entry (but in any case you can move it with the arrow keys).

8. Check files. This causes a window to appear displaying the number and date of the version of Leonora in use, followed by lines indicating whether the files LEONORA.MNU, LEONORA.ALT, and LEONORA.STR (which contain menu data, warning strings, and other strings respectively) have been altered since the program was installed.

9. Escape from program. Define whether pressing the ESCAPE key at the MAIN MENU allows exit from Leonora. In the FORBIDDEN state you can exit Leonora only by selecting Exit in the MAIN MENU (or pressing ALT-F9 in any menu), and if the data are not saved this will require confirmation. In the ALLOWED state you can exit Leonora (without a check of whether the data are saved) by pressing ESCAPE in the MAIN MENU.

10. Exit. Return to the MAIN MENU.

10

Customizing Leonora

10.1 Introduction and warning

This chapter is not for the casual user of Leonora, as the information here will allow you to render the program completely unusable if you do not proceed with due care and attention. In particular, do not carry out any of the changes discussed here without first checking that you have back-up copies of the following files: EQUATION.DAT, LEONORA.ALT, LEONORA.MNU, LEONORA.STR. All of these should be in the PRIVATE directory. You may find it advisable to copy all of the files in this directory to a floppy disk before starting. When Leonora is initially installed, all of these files, together with all the help files, are stored as hidden and read-only. This decreases the danger of accidentally destroying or corrupting them, but it has the inconvenience that they become invisible to DOS commands like DIR. Their file attributes can easily be changed by means of commercial utilities such as PC Tools. However, you can also do it with Leonora itself, by launching it with the command *LEONORA REVEAL*. This allows you to remove the hidden and read-only attributes from any files in the PRIVATE and HELPFILE directories (or reinstate them if you have previously removed them).

This said, Leonora is a highly customizable program, and it is possible to alter most of the menus and other information to your requirements, for example to delete functions that you do not use, to alter the key letters, to change to another language, etc. One way to do this is to edit the files mentioned above with a text editor, but this requires care as they contain control codes etc. whose function may not be obvious. A safer way is to use Leonora itself to edit these files, which is possible as long as you 'warn' it that you plan to do some editing: this requires that you press \ (back slash) while in the MAIN MENU. If you do not do this, any attempt to open the menu files will be ignored.

10.2 Editing a menu

While in the MAIN MENU, press \ (back slash). This will have no apparent effect, and all the normal functions of Leonora will remain in force. However, in addition, the entry ! (exclamation point) will be accepted as a valid option in any menu, and causes the menu to be opened for editing. To illustrate this, select any set of data, then open the OUTPUT REQUIRE-MENTS MENU, and press !. This will cause a new window to appear in which all of the relevant lines of the menu file are visible, including control characters that would ordinarily be invisible, i.e.

```
OUTPUT FILE NEEDED

\No output filePtNo
\ResultsPtSave resu
\DetailsPtSave resu
E\xitPtReturn to Ma
```

The special character Pt (ALT-158)* separates the menu entry from any help information that may be displayed on the bottom line of the screen. This part of the line (including any part that is initially missing from the display) can be edited in the same way as the menu entry itself, but in the discussion below it is ignored. The second line is blank because the menu in question has no information at the bottom of the frame. You can now choose any line with the arrow keys and edit it. Note that regardless of how you may have set the default behaviour of the Editor (Section 9.8), the cursor will be placed initially at the beginning of the line. For example, edit it to read as follows:

FICHIER DE RESULTATS

```
N\\\Aucun fichierPt ...
\RésultatsPt ...
!\DétailsPt ...
X\\\Sortie
```

If you now press ENTER and then select **Output** requirements again, you will find the menu appears in French and the entry **Details** is omitted (and the key letter **D** is not accepted).

Note that if you leave the key letter unchanged, as in the case of Résultats, it is sufficient just to replace the word by what you want, but if you change it, as in the case of Aucun fichier, you must type the original key letter (*N*) at the beginning of the line, followed by \\, and the new key letter must be preceded by \. An entry can be suppressed either by using the backspace key to delete it completely or by typing ! at the beginning. The latter is more convenient if you want to reinstate it later.

---

*This character may be displayed differently according to the version of DOS installed in your computer. On a standard French version of DOS it appears as Pt (peseta). The way it is handled by Leonora is determined by its index in the character set, i.e. 158, not by its appearance on the screen.

If you want to insert a new line, you must first press PAGE-DOWN or PAGE-UP while the cursor is in the line above or below (respectively) where you want the new line. Apart from wanting to reinstate a line that you have previously deleted, the main reason for wanting to do this is to define a macro, i.e. a series of commands that will execute automatically. For example, you might want to insert a new command **Calculate** in the DATA MENU that allows you to embark on a calculation immediately after you have read or entered a set of data. To do this, open the DATA MENU for editing and move the cursor to the line \Redefine data type and press SHIFT-TAB. Move to the next line and type

$\backslash$*Calculateμxcc*

If you do not have a μ on your keyboard you can type it as ALT-230. Everything before the μ defines the menu entry (make sure you do not choose a key letter that is already in use); everything after the μ defines the key strokes that would produce the same result.

After you have made any changes to any menu, these will remain active as long as you continue in Leonora. When you try to exit, a window will appear asking whether you want to restore the old file (saving the new one as NEWFILE.MNU) or accept the changes for use in the future (saving the old version as OLDFILE.MNU).

The menu file can also be edited manually with a text editor, and this allows the possibility of reordering the menus, which cannot be done from within Leonora. Provided that your computer can read rapidly from the hard disk there is no particular reason to do this, but if you have an older machine that takes an appreciable time to find a menu in the file (use F4 to check this, see Section 8.10), you may find it useful to bring frequently used menus to the beginning of the file. In order to do this, you should note that each menu is identified by a keyword such as mAINMENU that begins with a lower-case letter and continues in capitals and is terminated by the keyword eNDMENU. To move a menu, therefore, you should move all the text from and including the identifier line up to and including the terminator line. After such a move all menus should behave exactly as before apart from slight differences in the time required to find them.

Within a menu, you should take great care before altering any line consisting of a keyword such as eITHER, oR, eND, gET, or mAKE, or any line that follows a gET line. A series of lines that begins eITHER should always contain one oR line and be terminated with an eND line. A series beginning mAKE should be terminated with an eND line. A series beginning gET does not require a terminator line because it always contains exactly one line after the gET line. eITHER...oR...eND series can be nested within one another as if they were brackets { | }; thus in an eITHER...eITHER...oR...eND...oR...eND series the first oR and eND match the second eITHER, and the second oR and eND match the first eITHER, as in { { | } | }.

Initial and final spaces in menu lines are ignored by Leonora (and are used in the file for indenting series of lines to make the control structure more obvious). Fixed spaces are represented in the file by _ (underline).

10.3   Editing a warning

While in the MAIN MENU, press \ (back-slash). This will have no apparent effect, and all the normal functions of Leonora will remain in force. However, if you now press ! while a warning is on the screen, a cursor will appear and you will be able to edit the warning. After you have made any changes to any warning, these will remain active as long as you continue in Leonora. When you try to exit, a window will appear asking whether you want to restore the old file (saving the new one as NEWFILE.ALT) or accept the changes for use in the future (saving the old version as OLDFILE.ALT).

10.4   Editing other messages

Not all of the strings used by Leonora are stored in the files LEONORA.MNU and LEONORA.ALT. Others may be found in the file LEONORA.STR, or you may write your own file LEONORA.WRD to supersede what is in the program.

For example, in the file LEONORA.STR you will find the line

```
Weights calculated for simple errors
```

which is displayed if you have selected **Uniform** weights in the WEIGHTING MENU. If you prefer this to read

```
Uniform weights
```

you can edit the line accordingly, taking care not to alter anything else. In particular, do not alter any line that begins with a lower-case letter and continues in capitals, such as the line

```
sIMPLE
```

that precedes the line considered. If you do, for example if you change it to

```
Simple
```

Leonora will display

```
sIMPLE
```

when you expect it to display

```
Weights calculated for simple errors
```

This warning applies to all of the files in the PRIVATE directory. In the file LEONORA.MNU you will find many lines like

```
eITHER

oR

eND
```

etc. All of these are codes and must not be altered (Section 10.2). The words pH, pK, cAMP, mRNA, cDNA (and similar terms derived from RNA or DNA) are explicit exceptions: if you want to use them in a PRIVATE file you can do so without fear of confusing Leonora. Otherwise, do not write any lines that start with a lower-case letter and continue in capitals. Also, do not alter any line that *follows* a line reading

```
gET
```

If you cannot find the message you want to edit in the file LEONORA.STR, you may be able to edit it by adding a line showing the message exactly as it is displayed, and following it with a line containing the message you want displayed.

In many cases Leonora ignores blank characters at the beginning or end of a string read from one of these files. To type a fixed space, type _ (underline) instead.

You may be able to modify individual words (e.g. the word column that appears in the Input new data window), by writing a file LEONORA.WRD and storing it in the PRIVATE directory. For each word that you want to edit, first type a line showing how Leonora normally displays it, then a line showing how you want it, for example

```
column

colonne
```

Note that for this purpose column and columns are two different words, and to change both you must include two lines for each in the file LEONORA.WRD.

In all of these files the individual entries are identified by their names and not by their positions in the files. The order of entries can therefore be changed without affecting the results. This may be done to decrease access time by putting frequently used entries near the beginning and rarely used ones near the end.

10.5 Editing help files

All help files are stored in the HELPFILE directory and have the extension HLP. They cannot be edited from within Leonora, but are all text files and can be edited without danger of crashing Leonora by means of any text editor.

First determine the name of the help file by pressing **?** while the help screen is visible. This will display, for example, Filename: Leonora.hlp. Then exit from Leonora, open the file with a text editor and make whatever changes you want. In help files, the characters \, « (ALT-174),* and » (ALT-175)* are control characters and are not counted in the line length, which should not exceed 39. \ causes the character that follows to be highlighted. « causes everything that follows to be highlighted until the end of the line or ». A help file (other than QUICKY.HLP) should not contain more than 150 lines. Any excess lines are ignored.

QUICKY.HLP is somewhat different from other help files. Each page begins with a code beginning with a lower-case letter but otherwise in capitals, such as eQNFIRST, and ends with eNDHELP. These lines must not be altered. The message itself, which should not contain more than 13 lines, can be edited.

10.6 Editing the equation file

The main functions are available without opening the file EQUATION.DAT and are described in Section 9.3. However, there are three functions for which it may be more convenient to work directly on this file using a text editor, and one that can *only* be done by editing it directly.

1. *Simple editing*. To change the title of an equation or the name of a parameter, find the line in EQUATION.DAT containing the information that you want to change and edit it.

2. *Cleaning up the file*. If you frequently add equations and subsequently delete them, the file EQUATION.DAT will eventually contain many lines that are no longer in use. To remove these, find each line that begins XXXX; then delete this line, the previous line, and the following line.

---

* ALT-174 and ALT-175 may code for other characters in some versions of DOS. In any case it is the indexes in the character set that determine how they are handled by Leonora, not their appearance on the screen.

3. *Reinstating disabled entries.* Find the appropriate entry (recognizable by its title and by the fact that its second line begins XXXX and may end (entry deleted). Then delete XXXX and the space that follows XXXX.

4. *Treating parameters with different names as identical.* A parameter that occurs in two different equations may need to be treated as the same parameter in statistical calculations even though different symbols are used; for example Ks in the equation for substrate inhibition (Section 9.3.2, paragraph 1(b)) corresponds to Km in the Michaelis–Menten equation (Section 9.3.2, paragraph 1(a)). To indicate this, replace Ks in the equation file by *KsáKm* (á is ALT-160). This particular example is already implemented in the equation file when installed.

The following characters* have special meanings in EQUATION.DAT and should not be altered except in a spirit of experimentation: \, « (ALT-174), » (ALT-175), any number contained between « and », { (ALT-123), } (ALT-125), á (ALT-160), and à (ALT-133). This last is read by Leonora as (app.), i.e. (apparent).

---

* All of them (apart, probably, from \) may be displayed quite differently by different versions of DOS. In all cases it is the indexes in the character set that determine how they are handled by Leonora, not their appearance on the screen.

11
Use of simulated data

11.1 Introduction: generation of pseudo-random numbers

The sequence does not of course in reality consist of independent elements: on the contrary, for particular values of the constants, each element uniquely defines the next. But often the nature of the dependence is such that the usual statistical tests cannot detect it, and to all appearances, the values are indeed independent.

Neave 1973

There are two fundamentally different approaches for assessing the performance of different methods of fitting data. The first is to study theoretical statistics to the point where one can deduce analytically how to arrive closest to minimum-variance or maximum-likelihood estimates of the parameters of interest. This approach is, however, beyond the reach of most experimentalists who are more interested in analysing their data than in becoming statisticians. Fortunately there is a less rigorous alternative that is much easier to apply, namely to compare methods by observation of how they perform in practice. This cannot be done adequately with data from real experiments, because in real experiments one never knows what the true solution is and there is no way to judge which of two or more estimation procedures comes closest to it. Likewise one never knows in a real experiment how the errors are distributed. Even if these objections did not apply, real experiments would be of limited value because one could rarely if ever accumulate enough of them to give a statistically useful sample of results for comparison.

None of these difficulties applies to experiments simulated in the computer, however. One can define 'true' values at will, and add errors drawn from known distributions to generate an unlimited series of experiments. Provided that one takes reasonable precautions the procedure is straightforward and reliable. Virtually all computers provide functions for generating 'pseudo-random' numbers: this means that they are calculated from a function, and hence are completely predetermined and not really random at all, but none the less satisfy tests of randomness that do not take account of the function from which they are calculated. One advantage of this over a truly random procedure, such as counting radioactive disintegrations, is that it is very fast; in addition, it allows one to use the same numbers again (for example to repeat the same simulation with methods that were not included originally), provided one keeps a record of the starting value or 'seed' used to initialize the random number generating function.

The simplest and most widely used function is called a *linear congruential generator*, and is defined as follows:

$$x_{i+1} = (px_i + q) \bmod r \qquad (11.1)$$

in which x_i, p, q, and r are integers, i.e. each number is calculated from the previous number by multiplying by p, adding q and taking the remainder after dividing by r. As the resulting values of x_i are uniformly distributed integers in the range 0 to $r - 1$, it is usually convenient to convert them to real numbers u_i uniformly distributed in the range 0 to 1 (strictly 0 to $1 - 1/r$) by dividing each by r:

$$u_i = x_i/r. \qquad (11.2)$$

This type of generator can give very good results, but it can also give very bad results if implemented inappropriately, and for this reason it is useful to have some idea of how it works. In practice it is best to make r as large as possible within the constraints of the integer capabilities of the computer, and for reasons related to the way computers work it is usually an exact power of 2. However, the behaviour of the function is easier to understand if r is small enough for hand calculation and the entire series short enough to be examined. For illustration, therefore, it is convenient to take $r = 64$.

If $p = 13$, $q = 25$, and the starting value is 0, it is easy to show that the series is (0, 25, 30, 31, 44, 21, 42, 59, 24, 17, 54, 23, 4, 13, 2, 51, 48, 9, 14, 15, 28, 5, 26, 43, 8, 1, 38, 7, 52, 61, 50, 35, 32, 57, 62, 63, 12, 53, 10, 27, 56, 49, 22, 55, 36, 45, 34, 19, 16, 41, 46, 47, 60, 37, 58, 11, 40, 33, 6, 39, 20, 29, 18, 3, 0...). This then repeats indefinitely with a cycle of 64, so that $x_{i+64} = x_i$. At first sight this looks as random a way of arranging the integers from 0 to 63 as any other, but the more one studies it the more fault one can find. The most obvious point is that the numbers are alternately odd and even, and once this is noticed it does not take long to notice that every fourth value is divisible by four, and every eighth value by eight. If one expresses the series as binary numbers it becomes obvious that the less significant half of each number (i.e. the last three binary digits) is completely regular: (000000, 011001, 011110, 011111, 101100, 010101, 101010, 111011, 011000, 010001, 110110, 010111, 000100, 001101, 000010, 110011...). If one separates each number into more significant and less significant halves and converts backs into decimal notation, the more significant halves produce a series that is not noticeably

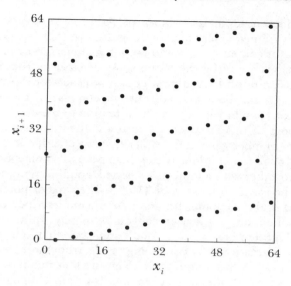

Fig. 11.1. *Lack of independence of consecutive 'random' numbers.* Each number x_i was converted into the next one x_{i+1} by the formula

$$x_{i+1} = (13x_i + 25)\bmod 64.$$

Although this is a deliberately bad choice of parameters for a random number generator, the problems evident in the figure occur to some extent with all numbers generated by linear congruential generators.

less random than the complete numbers: $(0, 3, 3, 3, 5, 2, 5, 7, 3, 1, 6, 1, 0, 2, 0, 1, 0, 6 \dots)$, but the series obtained from the less significant halves is hopelessly non-random $(0, 1, 6, 7, 4, 5, 2, 3, 0, 1, 6, 7, 4, 5, 2, 3, 0, 1 \dots)$. This characteristic persists when one uses much larger values for r than 64, and so it is important to resist the temptation to break up a pseudo-random number into supposedly random digits. Even if you only need a series of random switches that can be either 0 or 1, each one should be calculated separately as the most significant bit of a complete number, the rest of the number being discarded.

Plotting x_{i+1} against x_i (Fig. 11.1) reveals an even more serious problem with the series. While it would have been over-optimistic to expect to get a random scatter of points from such a simple calculation, one might still have hoped for something a bit better than a set of five straight lines. As is obvious from the points, the values of consecutive elements in the series are correlated, with a correlation coefficient of 0.181, much further from zero than one could regard as acceptable even if there were no particular need to use values in pairs. In practice, however, one often does need to use values in pairs, as a way of converting uniformly distributed values into more useful distributions (Section 11.2): it is essential for consecutive pseudo-random numbers to be uncorrelated if such conversions are to be valid, and gross departures from the expected distributions can result from a poor choice of generator (Neave 1973).

The problems can be alleviated (but not eliminated) by using a much larger value of r, and by giving careful attention to the choice of p and q. As the purpose of this discussion is to draw attention to some problems with random number generators rather than to describe in detail how to design one, I shall not continue with it here beyond commenting that fuller discussion, together with a table of suggested 'good' combinations of p, q, and r, may be found in Press *et al.* (1988). Their major points, which are implemented in the computer program SimuLeon described below, is that in serious simulation work it is dangerous to rely on any single linear congruential generator, even a 'good' one, and that one can improve performance dramatically by combining results from more than one generator and by using a shuffling routine to ensure that numbers are not used in the order in which they are generated. They also emphasize the point made above that one should *never* use the low-order half of a pseudo-random number as a random number in its own right.

11.2　Changing the distribution of pseudo-random numbers

Most random number generators generate uniformly distributed random numbers, i.e. all values within the allowed range are equally likely. This, however, is not at all how real experimental errors are distributed, and so for serious simulation we need a means of converting a uniform distribution into a normal distribution (if we accept the classical assumptions about error distribution, see Chapter 4) or other distributions (if we do not). Fortunately, there exists a simple formula, obtained by Box and Muller (1958), for converting a pair of uniformly distributed random numbers (u_1, u_2) in the range 0 to 1 into a pair of normally distributed random numbers (x_1, x_2) with mean 0 and standard deviation 1:

$$x_1 = (-2 \ln u_1)^{1/2} \cos 2\pi u_2, \tag{11.3}$$

$$x_2 = (-2 \ln u_1)^{1/2} \sin 2\pi u_2. \tag{11.4}$$

Perhaps surprisingly, x_1 and x_2 are statistically independent even though they are calculated from the same pair of values. As it is computationally more efficient to use both (rather than repeating the whole calculation for the second) even though one may only want one at a time, the usual practice is to compute both x_1 and x_2 on the first call, using x_1 and saving x_2 for the next call.

Marsaglia and Bray (1964) described a variant of the Box–Muller method that is computationally more efficient (because it avoids explicit calculation of the trigonometric functions) and is also somewhat more resistant to the numerical problems that can arise if the initial uniform random number generator is ill designed. In this method, u_1 and u_2 are first converted to uniformly distributed numbers $U_1 = 2u_1 - 1$ and $U_2 = 2u_2 - 1$ in the range -1 to 1. If $S = U_1^2 + U_2^2 \geq 1$ then u_1 and u_2 are discarded and a new pair generated, but if $S < 1$ then a pair of values x_1 and x_2 calculated as follows is independently normally distributed with mean 0 and standard deviation 1:

$$x_1 = U_1 \left(\frac{-2 \ln S}{S} \right)^{1/2} , \qquad (11.5)$$

$$x_2 = U_2 \left(\frac{-2 \ln S}{S} \right)^{1/2} . \qquad (11.6)$$

The relationship of this calculation to that of Box and Muller (1958) is explained by Press *et al.* (1988).

It is important to realize that eqns 11.3–11.6 are valid for uniformly distributed random numbers: this says nothing about applying them to pseudo-random numbers. In practice they work satisfactorily as long as one takes care to use a good random number generator and shuffles the values to eliminate residual correlations between successive numbers.

In practice we are unlikely to want a normally distributed random number with mean 0 and standard deviation 1, but fortunately it is very easy to convert to any desired mean and standard deviation: if x has mean 0 and standard deviation 1, then $\mu + \sigma x$ has mean μ and standard deviation σ. (This conversion applies regardless of whether x is normally distributed or not.)

Generating numbers from mildly long-tailed distributions is also quite easy as long as these are of 'contaminated normal' type, which means that they consist of a normal distribution contaminated by a small proportion of values from a different normal distribution. For example the middle panel of Fig. 4.2 showed such a distribution, in which each value had probability 0.95 of coming from a distribution with mean 10 and standard deviation 1, and probability 0.05 of coming from one with mean 10 and standard deviation 3. To obtain a number from this distribution, generate three uniformly distributed numbers u_1, u_2, and u_3; insert u_1 and u_2 into eqn 11.3 to calculate x_1; take $10 + x_1$ as the final number if $u_3 \leq 0.95$ but take $10 + 3x_1$ instead if $u_3 > 0.95$.

Generation of values from more exotic distributions will not be discussed here, apart from noting that the Cauchy distribution is even simpler to generate than the normal distribution, as it can be obtained as the ratio of the expressions in eqns 11.3 and 11.4:

$$x = \tan 2\pi u. \tag{11.7}$$

This gives a value x with a median of 0 and median deviation 1 (but with mean undefined and variance infinite). The Cauchy distribution is so grotesquely non-normal that it is not useful for realistic simulations, but it is valuable for producing examples to convince inveterate believers in classical methods that there can be circumstances where the median remains stable even though the mean fails completely.

11.3   Simulating Leonora

SimuLeon is a separate program included with Leonora that allows study of the behaviour of Leonora in simulated experiments with observations of known error structure. Leonora itself contains a simulation function (Section 9.2.1), but for several reasons this is not appropriate for long-term repetitive use: the sequence of keys to be pressed is not very convenient; the results are not displayed in a way that facilitates comparison between numerous experiments; most important, perhaps, Leonora uses the ordinary random number generator built into the computer, which has insufficiently reliable statistical properties for serious simulations, as discussed above (Section 11.1).

The description in this chapter of how to use SimuLeon is less detailed than the description of Leonora in Chapters 8–10, because it is written with two assumptions: first, users of SimuLeon will already be familiar with the menu structure and functions of Leonora, so that features that work the same way in both programs do not need to be laboured; second, such users are likely to be more expert in general about running computer programs. The major emphasis, therefore, will be to explain features in which SimuLeon behaves differently from Leonora.

In the same spirit SimuLeon does not have associated help files: pressing F1 brings up a message saying that no help is available. Similarly, it is less concerned to trap user errors than Leonora: for example, if you try to use a non-existent data file, or start the simulation with no equations or methods defined, it will crash rather than give an error message.

11.4   Entering data

SimuLeon contains no data management functions—no data menu, no input of data, no editing, not even the facility to choose an existing data file from within the program. This means, first of all, that it cannot be used independently of Leonora (unless you go to enormous and pointless effort), and that it should reside in the same directory as Leonora. If you launch it with the DOS command *SIMULEON* ENTER, it will use the last set of data used with Leonora, if this still exists, or will crash if it does not. Alternatively you can name the data file in the DOS command, for example *SIMULEON EXAMPLE1.MMD* ENTER. You can omit the extension if it is *.MMD*.

SimuLeon uses the values of the independent variables in the data file for deciding the experimental design; values of the dependent variable are ignored. Although new values are simulated during computation these are not stored, and SimuLeon leaves the data file exactly as it finds it.

11.5 Selecting equations

SimuLeon contains two EQUATIONS MENUS, one for deciding the true equation, i.e. the one to be used in simulating the data, and the other for deciding which equations are to be fitted. The first works exactly like the corresponding menu of Leonora (Section 9.3) except that there are no functions for editing the equation file. The second is similar to the EQUATIONS MENU in Leonora, but allows you to select up to four equations to be fitted in sequence (in the same order that they are selected). However, if you have already selected more than one method (Section 11.7) you can only select one equation, i.e. SimuLeon allows you to compare methods or to compare equations, but not both at the same time.

In the MAIN MENU the key letters corresponding to the equations currently selected are shown after the two equations entries.

11.6 True parameter values, error parameters, output

The TRUE PARAMETER VALUES MENU prompts you for the value of each parameter in the true equation, to be used in generating the data.*

The ERROR PARAMETERS MENU works exactly like the corresponding Leonora menu (Section 9.2.1, paragraph 10).

The OUTPUT REQUIREMENTS MENU is similar to the corresponding Leonora menu (Section 9.4), except that the results are written to a file DATAFILE.SIM (for a data file called DATAFILE.MMD), and the interpretation of **Details** is different; with this option the first ten sets of data (and not just the parameter values) are saved; after the first ten cycles there is no difference between **Results** and **Details**.

11.7 Methods and weights

The METHODS MENU behaves as in Leonora (Section 9.5.1), except that you

---

* If this step is omitted, which is permissible but in general is not advisable, each parameter value is set to 1, not to values relevant to the original data on which the design is based.

can select up to four different methods, to be used in the order they are selected, provided that you have not selected more than one equation to be fitted (Section 11.5), in which case only one method can be selected.

The WEIGHTS MENU cannot be used independently, but appears automatically if you select a method that requires weights, i.e. least squares or least absolutes. It then behaves as in Leonora (Section 9.5.2). Each method-and-weights combination counts as one method in the total of four allowed, and if you select the same combination twice it counts as two. On returning to the MAIN MENU the number of methods selected appears after Methods. Initially this is 1, because if no method is explicitly selected the Leonora default method is used (Section 9.5.1).

11.8   Results screen

Once equations, methods, and error function have been defined, the simulation can be launched with Start simulation. This then continues repetitively until ESCAPE is pressed (up to a maximum of 32 767 repetitions). The information displayed varies according to whether several equations are being fitted by the same method or several methods are being used with the same equation, or one method is being used with one equation but the data contain replicates, but some general points will be noted first.

As the different modes of displaying results may seem arbitrary, it may be useful to spell out the general principle that decides what choices SimuLeon allows you to make, and what it does with the data presented to it: SimuLeon assumes that you only want to make one sort of study at a time (and whether this is what you want or not, it is all SimuLeon will allow). If you choose more than one method, therefore, SimuLeon assumes you are comparing methods and not equations; if replicates are present in the data and just one method is being used to fit one equation, it uses the replicates to do an analysis of variance for lack of fit and pure error, but if there is more than one method or more than one equation, the replicates are treated as ordinary observations. This means of course, that you may need to run the program several times with the same data to obtain all the results you need.

In all modes lines 1 and 25 are the same as when a menu is on the screen, but *note* that an asterisk * in the middle of line 1 has a new meaning from that in Leonora: in Leonora this means that the data file is not saved (Section 8.9), but this would have no point in SimuLeon because one cannot work with a data file that has not been saved; instead, therefore, it indicates that the results are not being written to a file.

If not all previous calculations are still in memory, the first calculation still available is indicated at the end of line 2 (after whatever other information is described in the sections below). You can increase the amount of previous results that SimuLeon is able to remember by running it without other programs resident in memory simultaneously.

As a simulation may continue for several hours, the screen remains bright for only the first 5 min. After this time it switches to a faint display against a dark background, and after a further 5 min it switches to black on black (i.e. invisible to the observer, though the calculation continues). In either of these states the normal display can be recovered by pressing any key (after a delay until the current calculation is finished). Although the Press any key... message that appears with the black-on-black display is superficially similar to the message that may appear when SimuLeon (or Leonora) is waiting for a menu selection, it does not imply that the computer is idling.

During the normal display the following keys are active: PAGE-DOWN causes more lines to be devoted to individual simulations and fewer to the summaries, and PAGE-UP has the opposite effect (i.e. they move the dividing line between the upper and lower parts of the display down and up respectively). If fewer summary panels are currently displayed than the number that exist, the arrow keys ↑ ↓ allow scrolling backwards or forward between the possible panels; other arrow keys and function keys have no effect; the ESCAPE key terminates the simulation (*note* no warning!); any other key switches the display to recapitulation mode (described in the next paragraph).

In recapitulation mode the simulation pauses and the whole screen is devoted to the individual simulations done so far, and one can scroll back to the first line still in memory by means of the arrow keys. The selected line is marked by ■ (for the line corresponding to the first set of parameter values) or · (for other lines) in column 1. When ■ is displayed the TAB key becomes active, and causes a window to appear listing the series of values of the dependent variable corresponding to the selected simulation. Pressing any key removes this window, and pressing any key again causes simulation to resume.

11.8.1 One equation, one method

Line 2. This is a caption with the parameter names.

Lines 3–18. The results of the calculation are added to these lines such that each new line is added at the bottom and earlier lines scroll off the top. Each calculation generates a line showing the linear parameter values and the ordinary parameter values, and the cycle number is shown. If the method is least squares with dynamic weights, the approximate value of σ_0/σ_2 or α (Section 9.5.2) is indicated by the following code: R, relative or nearly relative weights ($\alpha > 1.9$); r, tending towards relative weights

$(1.3 < \alpha < 1.9)$; I, near the middle of the range between relative and uniform weights $(0.7 < \alpha < 1.3)$; u, tending towards uniform weights $(0.1 < \alpha < 0.7)$; U, uniform or nearly uniform weights $(\alpha < 0.1)$.

Line 19. This is a caption showing the method used.

Lines 20–3. These show respectively the maximum, minimum, mean, and standard deviation of each parameter.

Line 24. This shows (if the method generates confidence limits) the proportion of times that each true linear parameter value was within the calculated limits.

11.8.2   Two equations

Line 2. This shows a caption indicating the method used. If this is least squares with dynamic weights, it switches to fixed weights for the second equation (Section 9.5.2).

Lines 3–7. The first two lines for each set of data are set out as lines 3–18 in Section 11.8.1. If the method is least squares, there is also a third line that shows an analysis of variance for the improvement due to the additional parameter.

Lines 8–13 and 14–19. These show summary panels for the first and second equations, with captions on lines 8 and 14 showing the parameter names and the rest as for lines 20–3 in Section 11.8.1.

Lines 20–4. These summarize the results of the analyses of variance, with the following columns: residual mean square (for the equation with most parameters), mean square (for the additional parameter), value of F in the significance test, and the logarithm of the corresponding probability. The right-hand part of the panel shows the proportions of trials in which various levels of significance were attained.

If the method in use is not least squares, the bottom panel is omitted (i.e. there is no analysis of variance) and the additional lines available are used to display more results of individual experiments.

11.8.3   More than two equations

The display is as in Section 11.8.2 except that there is no summary of the results of analysis of variance (lines 20–4 in Section 11.8.2). Models are compared by analysis of variance only when there are exactly two equations being fitted: to study larger numbers of models, run them two at a time (see Section 11.9 for information about how to set the randomizer to regenerate the same random numbers in subsequent runs).

11.8.4   More than one method

Line 2 shows parameter names and the rest is divided into a results panel and one to three method summaries essentially as above. As there may be more methods than there are summary panels, one can scroll between panels using the arrow keys.

11.8.5   Replicate observations

If there is more than one equation or more than one method, replicate observations in the data are given no special treatment: each member of a group of replicates is treated like any other observation. However, if there is only one method, and this is least squares, and only one equation, then the replicate observations are used to estimate the sum of squares for pure error and carry out an analysis of variance for lack of fit (Section 9.5). The raw results appear as lines added to the result lines at the top of the screen, and a special panel showing maximum, minimum, mean, and standard deviation values appears at the bottom of the screen.

11.9   Randomizer

Normally SimuLeon uses the internal clock to initialize the randomizer, so that different random numbers are used in each run. To initialize with a particular value (in the range 0 to 32 767) select this entry in the MAIN MENU and then enter the value required after selecting **New seed**. To restart using the same sequence of random numbers used already in the same session (i.e. without exiting from SimuLeon), select **Reinitialize with . . . .** If you note on a piece of paper the number that appears after **Reinitialize with . . .** you can use it in a future session to regenerate the same numbers. *Note*, however, that this number must be exactly the same: 12 345 generates an entirely different sequence from 12 344, for example. Moreover, the same random numbers will only generate the same data (even approximately) if the number of observations and the error function are unchanged.

In contrast to the simulation function in Leonora, SimuLeon generates uniformly distributed random numbers by the function RAN1 of Press *et al.* (1988), which uses three linear congruential generators simultaneously, the first for the more significant part of the output number, the second for the less significant part, and the third to shuffle each group of 97 numbers into an arbitrary order (cf. Section 11.2). The resulting uniformly distributed numbers are then used for generating numbers from a normal (Gaussian) distribution by the method of Marsaglia and Bray (1964).

References

Anscombe, F. J. (1960). *Technometrics*, **2**, 123–47.

Anscombe, F. J. (1973). *American Statistician*, **27**, 17–21.

Askelöf, P., Korsfeldt, M., and Mannervik, B. (1976). *European Journal of Biochemistry*, **69**, 61–7.

Bowley, A. L. (1928). *F. Y. Edgeworth's contributions to mathematical statistics*, p. 101. Royal Statistical Society, London.

Box, G. E. P. and Muller, M. E. (1958). *Annals of Mathematical Statistics*, **29**, 610–11.

Burk, D. (1934). *Ergebnisse der Enzymforschung* **3**, 23–56.

Cárdenas, M. L. and Cornish-Bowden, A. (1993). *Biochemical Journal*, **292**, 37–40.

Chambers, J. M., Cleveland, W. S., Klein, B., and Tukey, P. A. (1983). *Graphical methods for data analysis*, p. 1. Wadsworth, Belmont, CA.

Cleland, W. W. (1967). *Advances in Enzymology*, **29**, 1–32.

Cleland, W. W. (1979). *Methods in Enzymology*, **63**, 103–38.

Colquhoun, D. (1973). *Lectures on biostatistics*. Clarendon Press, Oxford.

Cornish-Bowden, A. (1976). *Principles of enzyme kinetics*. Butterworths, London.

Cornish-Bowden, A. (1982). *Journal of Molecular Science (Wuhan)*, **2**, 107–12.

Cornish-Bowden, A. (1983). *Journal of Theoretical Biology*, **101**, 317–19.

Cornish-Bowden, A. (1985). In *Techniques in protein and enzyme biochemistry* (ed. K. F. Tipton), BS115, pp. 1–22. Elsevier, Limerick.

Cornish-Bowden, A. (1995). *Fundamentals of enzyme kinetics* (2nd edn). Portland Press, London.

Cornish-Bowden, A. and Cárdenas, M. L. (1987). *Journal of Theoretical Biology*, **124**, 1–23.

Cornish-Bowden, A. and Eisenthal, R. (1974). *Biochemical Journal*, **139**, 721–30.

Cornish-Bowden, A. and Eisenthal, R. (1978). *Biochimica et Biophysica Acta*, **523**, 268–72.

Cornish-Bowden, A. and Endrenyi, L. (1981). *Biochemical Journal*, **193**, 1005–8.

Cornish-Bowden, A. and Endrenyi, L. (1986a). *Biochemical Journal*, **234**, 21–9.

Cornish-Bowden, A. and Endrenyi, L. (1986b). In *Dynamics of biochemical systems* (ed. T. Keleti, S. Damjanovich, and L. Trón), pp. 309–22. Akadémiai Kiadó, Budapest.

Cornish-Bowden, A. and Wong, J. T. (1978). *Biochemical Journal*, **175**, 969–76.

Cornish-Bowden, A., Porter, W. R., and Trager, W. F. (1978). *Journal of Theoretical Biology*, **74**, 163–75.

Daniels, H. E. (1954). *Annals of Mathematical Statistics*, **25**, 499–513.

Draper, N. R. and Smith, H. (1981). *Applied regression analysis* (2nd edn). Wiley, New York.

Eisenthal, R. and Cornish-Bowden, A. (1974). *Biochemical Journal*, **139**, 715–20.

Endrenyi, L. (ed.) (1981). *Design and analysis of experiments in enzyme and pharmacokinetic experiments.* Plenum Press, New York.

Fisher, W. D. (1961). *Journal of the American Statistical Association*, **56**, 359–62.

Gauss, K. F. (1809). *Theoria motus corporus coelestium in sectionibus conicis solem ambientium* (trans. C. H. Davis), Section 177, pp. 257–9. Dover, New York, 1963 (republication of edition of 1857 by Little, Brown & Co.).

Geary, R. C. (1947). *Biometrika*, **34**, 209–42.

Haldane, J. B. S. (1948). *Biometrika*, **35**, 414–15.

Hojo, T. (1931). *Biometrika*, **23**, 315–60.

Johansen, G. and Lumry, R. (1961). *Comptes rendus des Travaux du Laboratoire Carlsberg*, **32**, 185–214.

Kendall, M. G. (1970). *Rank correlation methods*, (4th edn.). Griffin, London.

Kendall, M. G. and Stuart, A. (1969). *The advanced theory of statistics* (3rd edn). Vol. 1. Griffin, London.

Laplace, P. S. (1798). *Mécanique céleste*, Book 3, Chapter 5, Section 40. Crapelet, Paris. (*Œuvres complètes de Laplace*, Vol. 2, pp. 143–7. Gauthiers-Villars, Paris, 1878.)

Lineweaver, H., Burk, D., and Deming, W. E. (1934). *Journal of the American Chemical Society*, **56**, 225–30.

Mannervik, B., Jakobson, I., and Warholm, M. (1986). *Biochemical Journal*, **235**, 797–804.

Marsaglia, G. and Bray, T. A. (1964). *SIAM Review*, **6**, 260–4.

Michaelis, L. and Menten, M. L. (1913). *Biochemisches Zeitschrift*, **49**, 333–69.

Mosteller, F. and Tukey, J. W. (1977). *Data analysis and regression*, pp. 333–79. Addison-Wesley, Reading, MA.

Neave, H. R. (1973). *Applied Statistics*, **22**, 92–7.

Nelder, J. A. (1991). *Biometrics*, **47**, 1605–15. (includes an additional discussion by D. Ruppert, N. Cressie, R. J. Carroll, and J. A. Nelder).

Nimmo, I. A. and Mabood, S. F. (1979). *Analytical Biochemistry*, **94**, 265–9.

Philo, R. D. and Selwyn, M. J. (1973). *Biochemical Journal*, **135**, 525–30.

Poincaré, H. (1892). *Thermodynamique*, p. v. Georges Carré, Paris.

Press, W. H., Flannery, B. P., Teukolsky, S. A., and Vetterling, W. T. (1988). *Numerical recipes: the art of scientific computing*, pp. 192–203. Cambridge University Press.

Raaijmakers, J. G. W. (1987). *Biometrics*, **43**, 793–803.

Raphel, V. (1994). *Thèse du 3ᵉ cycle*. Université d'Aix-Marseille III.

Ruppert, D., Cressie, N., and Carroll, R. J. (1989). *Biometrics*, **45**, 637–56.

Sen, P. K. (1968). *Journal of the American Statistical Association*, **63**, 1379–89.

Siano, D. B., Zyskind, J. W., and Fromm, H. J. (1975). *Archives of Biochemistry and Biophysics*, **170**, 587–600.

Storer, A. C., Darlison, M. G., and Cornish-Bowden, A. (1975). *Biochemical Journal*, **151**, 361–7.

Tukey, J. W. (1962). *Annals of Mathematical Statistics*, **33**, 1–67.

Tukey, J. W. (1977). *Exploratory data analysis*. Addison-Wesley, Reading, MA.

Tukey, J. W. and McLaughlin, D. H. (1963). *Sankhyā Series A*, **25**, 331–52.

Watts, D. G. (1981). In *Design and analysis of experiments in enzyme and pharmacokinetic experiments* (ed. L. Endrenyi), pp. 1–24. Plenum Press, New York.

Wilkinson, G. N. (1961). *Biochemical Journal*, **80**, 324–32.

Index